Enterprise Mobility with App Management, Office 365, and Threat Mitigation
Beyond BYOD

Yuri Diogenes
Jeff Gilbert
Robert Mazzoli

PUBLISHED BY
Microsoft Press
A division of Microsoft Corporation
One Microsoft Way
Redmond, Washington 98052-6399

Library of Congress Control Number: 2015951523
ISBN: 978-1-5093-0133-1

Printed and bound in the United States of America.

First Printing

Microsoft Press books are available through booksellers and distributors worldwide. If you need support related to this book, email Microsoft Press Support at mspinput@microsoft.com. Please tell us what you think of this book at http://aka.ms/tellpress.

Acquisitions Editor: Karen Szall
Developmental Editor: Karen Szall
Editorial Production: Christian Holdener, S4Carlisle Publishing Services
Technical Reviewer: Mike Toot; Technical Review services provided by Content Master, a member of CM Group, Ltd.
Copyeditor: Roger LeBlanc
Indexer: Maureen Johnson, MoJo's Indexing
Cover: Twist Creative • Seattle

Contents at a glance

Contents

What do you think of this book? We want to hear from you!

Microsoft is interested in hearing your feedback so we can improve our books and learning resources for
you. To participate in a brief survey, please visit:

http://aka.ms/tellpress

What do you think of this book? We want to hear from you!

Microsoft is interested in hearing your feedback so we can improve our books and learning resources for you. To participate in a brief survey, please visit:

http://aka.ms/tellpress

Foreword

I speak with hundreds of IT Pros and CIOs every year, and each of them has the same priority: providing their users with an iconic work environment while securing and protecting company data. Doing this has become more difficult than ever thanks to the combination of more apps/data moving to the cloud and cyberattacks becoming more destructive. It is safe to say that the traditional perimeter that was used in the past to protect company data has evaporated; this means that organizations need to fundamentally rethink how they are securing and protecting company data. Microsoft has committed itself to being an ally to the IT professionals charged with protecting the assets of their companies.

It is no exaggeration to say that, at Microsoft, we are obsessed with enterprise security. Every software company struggles with the balance between making corporate data safe from attack but accessible to the appropriate parts of the workforce—and I believe Microsoft has struck the right balance.

Teams across the company have torn down the traditional barriers that existed between products and built end-to-end solutions that are not just interoperable, but built to protect data wherever it goes. This means protecting it at multiple layers throughout the organization: protecting it at the device and apps (with Microsoft Intune), protecting the file (with Azure RMS), and protecting identities (with Azure Active Directory Premium and Advanced Thread Analytics). These products all come together to form the Enterprise Mobility Suite (EMS).

This book is written by a trio of EMS experts, and it offers an insider's look at proven, real-world actions you can take to manage your enterprise mobility needs, enable your workforce to be productive (across devices and platforms) with an iconic work experience, and help you protect your organization's assets and your workforce's privacy.

As you read, I think you'll be consistently impressed by the ways you can leverage EMS's powerful ability to deliver an incredible work experience for your users that correctly balances between user empowerment and data protection. To do this, we have engineered EMS and Office 365 to be used together.

The value and power of what we've built is widely recognized by the IT industry —EMS has already outgrown its competitors and continues to regularly add more features and functionality. We are committed to continuing to build, refine, and deliver the tools you need to protect your organization and empower it to do more.

Brad Anderson, Microsoft Corporate VP, Enterprise Client & Mobility
@InTheCloudMSFT

Introduction

Enterprise mobility management is one of the fastest-growing areas in the Information Technology field, and having a solid understanding of the newest features and capabilities is an important part of configuring and managing mobile devices. This book continues forward from the information covered in *Enterprise Mobility Suite: Managing BYOD and Company-Owned Devices* (Microsoft Press, 2015) and covers the fundamentals and capabilities of several Microsoft mobility management resources; the newest mobile application management features in Microsoft Intune, Microsoft Advanced Threat Analytics (ATA), and Mobile Device Management for Office 365 (MDM for Office 365). Throughout this book, we guide you through all the areas associated with planning, designing, and implementing these mobility management solutions.

Is this book for you? This book is for enterprise IT professionals who are responsible for implementing and managing mobility management technology as well as professionals charged with identifying and mitigating networking threats to on-premises networks. It is also meant to provide foundational expertise to IT professionals who aren't already familiar with these solutions or just want to learn more. We assume that the readers are familiar with the primary components of the Microsoft Enterprise Mobility Suite (EMS) and Office 365. It is also helpful to have basic knowledge about network-security principals and network-infrastructure components.

The scenarios described in this book are meant to be an end-to-end journey for each of the mobility management solution areas. They start with understanding overviews of each solution and then move on to implementing specific features and capabilities in the example organization. After completing the example scenarios, you'll have learned how to

- Manage and publish mobile applications, and deploy them to mobile devices and computers

- Deploy and configure the ATA Center and Gateway, including configuring reports to help identify suspicious activities

- Activate and configure MDM for Office 365, including enrolling and managing mobile devices

Acknowledgments

The authors would like to thank Karen Szall and the entire Microsoft Press team for their support in this project, Brad Anderson for writing the foreword of this book, and all the other Microsoft colleagues who contributed by reviewing this book: Gershon Levitz, Ophir Polotsky, Benny Lakunishok, Michael Dubinsky, Simon May, Sonia Wadhwa, Stacia Snapp, Owen Yen, Paul Mayfield, Joey Glocke, Rob Stack, and Karthika Raman. In addition:

Yuri   I would like to thank my wife and daughters for their support and understanding, my great God for giving me strength and continuing to guide my path, my friends and co-authors Jeff Gilbert and Robert Mazzoli (you guys rock!), the Microsoft ATA Team in Israel for the endless support on this project, and last but not least, my parents for working hard to give me an education, which is the foundation that I use every day to keep moving forward in my career.

Jeff   I would like to thank my wife, Chrissy, and kids, Nick, Haylee, Jackson, and Jillian for their love, patience, and unending support and encouragement throughout the long hours required to author content for a technical book of this depth. Also, thanks to my co-authors who kept me motivated to write and who I can count on every day for expert counsel and advice. Thanks also to the other Microsoft enterprise client management engineering team members who made themselves available for my never-ending stream of questions and clarifications.

Robert   I would like to thank my daughter, Alyssa, for inspiring me to follow in her footsteps and become a published author; Barbara for being the love of my life and for all her patience and understanding; Bruna and Luciano for the use of their kitchen table and all the wonderful Italian meals that fueled my writing; my co-authors Jeff Gilbert and Yuri Diogenes for their ongoing friendship, guidance, and motivation; and "all" my parents (Constance, Claude, Henri, and Kathy) for a lifetime of love and support.

Free ebooks from Microsoft Press

From technical overviews to in-depth information on special topics, the free ebooks from Microsoft Press cover a wide range of topics. These ebooks are available in PDF, EPUB, and Mobi for Kindle formats, ready for you to download at:

http://aka.ms/mspressfree

Check back often to see what is new!

Errata, updates, & book support

We've made every effort to ensure the accuracy of this book and its companion content. You can access updates to this book—in the form of a list of submitted errata and their related corrections—at:

http://aka.ms/EM2/errata

If you discover an error that is not already listed, please submit it to us at the same page.

If you need additional support, email Microsoft Press Book Support at *mspinput@microsoft.com*.

Please note that product support for Microsoft software and hardware is not offered through the previous addresses. For help with Microsoft software or hardware, go to *http://support.microsoft.com*.

We want to hear from you

At Microsoft Press, your satisfaction is our top priority, and your feedback our most valuable asset. Please tell us what you think of this book at:

http://aka.ms/tellpress

The survey is short, and we read every one of your comments and ideas. Thanks in advance for your input!

Stay in touch

Let's keep the conversation going! We're on Twitter: *http://twitter.com/MicrosoftPress*.

CHAPTER 1

Understanding Microsoft enterprise mobility solutions

Enterprise mobility management solutions aren't as simple anymore as connecting a few mobile devices to an email server or allowing some users to access company resources via a remote connection. Today's IT departments must support a much more robust and comprehensive user experience for modern employees. Users expect, and often even demand, application-feature and data-access parity between their mobile devices and the devices they use at the office. Add in the new challenges that IT departments face with managing cloud-computing services, user identity, applications, data security, and threat mitigation, and the enterprise mobility management landscape becomes much more complex and difficult to deploy and manage.

This chapter explains how Microsoft enterprise mobility solutions address these areas and covers the basics of enterprise mobility management. It also covers considerations for selecting and deploying these solutions, as well as introducing a sample enterprise mobility management scenario that will be used throughout this book.

Enterprise mobility management concepts

In enterprise IT management, companies are fully embracing the modern "work anywhere, from any device" vision. Trends like *bring your own device (BYOD)* and *mobile application management (MAM)* aren't just buzzwords or passing fads likely to fade out after a year or two. These concepts are part of the larger modern IT strategy supporting the consumerization of IT and the empowerment of users. Central to this strategy are cloud services, such as Microsoft Azure Active Directory and Microsoft Office 365. Leveraging the computing scale and ubiquity of access that these and other Software as a Service (SaaS) platforms provide to mobile devices and users requires planning and considering things from a different perspective than in the past.

Enterprise mobility management isn't just about connecting mobile devices to cloud services or resources. In fact, it's less about *devices* and more about *people*. Forward-looking organizations aim to empower employees and increase their productivity; the devices (mobile or not) they use are merely tools to help accomplish their work. This paradigm shift from a *device-centric* management structure to a *people-centric* management structure is significant. All the components that enable mobile productivity in an enterprise mobility management solution must have a people-centric architecture that aligns with enabling this vision. Finding the proper balance where employee empowerment and productivity meet the business needs of your organization is the crucial requirement for any enterprise mobility management solution.

With this vision in mind, be aware that a well-designed enterprise mobility management solution must address several key areas of the modern workplace, as shown in Figure 1-1.

- Users
- Devices
- Apps
- Data
- Protection

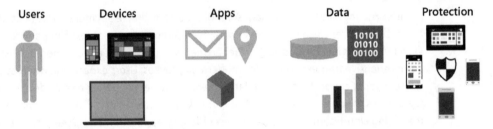

FIGURE 1-1 Elements of enterprise mobility management

> **MORE INFO** For more information about why organizations should embrace enterprise mobility solutions, read Chapter 1 of *Enterprise Mobility Suite: Managing BYOD and Company-Owned Devices* from Microsoft Press at *https://aka.ms/EMSdevice/details*.

Users

The first and most important element of the enterprise mobility management solution is the user or employee. Without the employee, the IT infrastructure and management costs to enable enterprise mobility are expensive monuments to best intentions. The enterprise mobility solution must support effective ways to manage user accounts and make it easy for employees to access resources. If user identity is hard to manage by IT administrators, or if employees are required to take convoluted steps to gain access to devices or company resources, the enterprise mobility management solution becomes an obstacle instead of an effective productivity management tool. As most experienced IT administrators have learned, workplace technology obstacles invite shortcuts, workarounds, and questionable data-protection practices.

Effectively managing user identity is critical to enabling cloud-based applications and data resources spanning multiple services or locations. Efficiently verifying that users are who they claim to be is essential to protecting resources and making the mobile experience feel like the traditional workplace experience. Keep in mind that employees with different types of roles and responsibilities, and even different geographic locations, often have unique requirements across all the areas of enterprise mobility management.

Devices

The rapid pace of technological advancement has changed the modern workplace from one of stationary workstations and company-issued devices to one containing a mix of all types of mobile computers and Internet-connected devices. This change is driving the BYOD trend across all markets, and industries and organizations must adapt to this new challenge. Using their personal mobile devices—such as smart phones, tablets, and laptops—employees are increasingly mixing their personal lives with their work responsibilities. As a result, IT departments are tasked with managing an ever-expanding collection of different mobile hardware, operating systems, and vendor-specific architectural requirements.

It's critical that organizations fully understand the capabilities and limitations of each type of device and how they will support each one. Only then can organizations define and configure the necessary enterprise mobility management features that support both the employee's needs and the organizations business requirements.

Apps

Apps are the centerpiece of most business requirements and the portal for information access for modern organizations. Though managing different device types creates new administration challenges, managing a mixture of commercial and customized line-of-business (LOB) apps can be equally challenging. Employees need access to all their productivity tools from all their devices, including email, data storage services, and role-specific tools. These services can be either locally hosted in on-premises networks or hosted in the cloud.

How to properly install and manage these apps depends on several factors. Different apps have different installation requirements, can require individual adjustments to function properly on different devices, and often have varying levels of risk associated with keeping information secure. Misjudging or improperly managing any of these areas can lead to exposing sensitive company data or employee personal information. IT departments must take care to fully understand which apps will be supported and how they will be managed to help protect company data. Mobile application management will be covered in more depth in Chapter 2, "Introducing mobile application management with Microsoft Intune," and Chapter 3, "Implementing MAM."

Data

Working from a mobile device from any location really means accessing data from anywhere. Operating hand in hand with identity management, apps, and the architecture of mobile devices, data must be consumed securely and easily for users to be productive and to keep them from finding alternative access routes to information. Understanding how data is stored on devices and how data is protected in transit is critical when planning and configuring enterprise mobility management features and policies.

Depending on your business needs and user requirements, your organization might require multiple layers of data protection, ways to classify information according to sensitivity, methods for data encryption, and integrated ways to manage access control. Different enterprise mobility management solutions offer varying levels of control for each of these areas and offer different levels of reporting and monitoring in the case of breaches.

Protection

Protecting mobile devices and company data from threats is just as important as securing data access. No matter how carefully planned security is, all levels of mobile device security are potentially vulnerable to a wide variety of malicious activity. These vulnerabilities include threats to company data, personal information, and even user identity.

Depending on the enterprise mobility management solution, preventing risk and protecting mobile devices from these threats can be included as tightly integrated features or standalone services. Understanding how these solutions address potential gaps in threat mitigation is extremely important to effectively protecting mobile devices that are coming from the cloud or located on-premises. Threat protection and mitigation will be covered in more depth in Chapter 4, "Introducing Microsoft Advanced Threat Analytics," and Chapter 5, "Implementing Microsoft Advanced Threat Analytics."

Microsoft enterprise mobility solutions

Microsoft has aggressively pursued a strategy of "mobile first, cloud first" in their enterprise mobility management vision. This vision is centered on helping organizations enable their users to be productive on the devices they prefer, while protecting company resources. Central to this vision is the concept of *balance*—balancing the financial and data-security needs of the company with the productivity and privacy needs of users. Finding an appropriate balance often means splitting authority between the company and users, and keeping added management complexity to a minimum to ensure satisfaction and compliance.

Instead of piecing together parts of existing on-premises products and attempting to update and rebrand them as cloud services, Microsoft chose to design an enterprise mobility management solution from the ground up and leverage the powerful features of its proven cloud services, such as Azure and Office 365.

Microsoft Enterprise Mobility Suite

The Enterprise Mobility Suite (EMS), shown in Figure 1-2, is a comprehensive set of cloud services and on-premises technologies designed to extend user identities to the cloud, manage mobile devices and apps, increase user productivity through native support for Microsoft Office apps and support for thousands of SaaS applications, and protect files accessed and stored on managed devices.

EMS comprises the following products:

- Microsoft Azure Active Directory Premium
- Microsoft Intune
- Microsoft Azure Rights Management
- Microsoft Advanced Threat Analytics

Identity Management	Microsoft Azure Active Directory Premium Cloud-based directory services and application access management
Mobile Device & Application Management	Microsoft Intune Cloud-based device configuration and management
Access & Information Protection	Microsoft Azure Rights Management Cloud-based data protection and data access management
Threat Protection and Mitigation	Microsoft Advanced Threat Analytics On-premises threat protection and threat notification

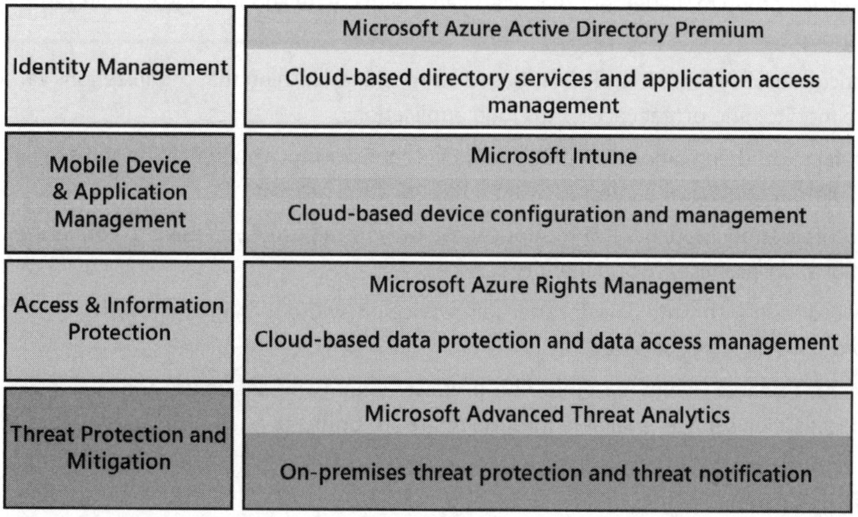

FIGURE 1-2 Enterprise Mobility Suite products

> **IMPORTANT** This book doesn't cover all the products included in EMS in depth. Instead, it focuses on several key features and capabilities of some EMS services, such as mobile application management (without device-enrollment requirements) and threat protection using Advanced Threat Analytics. It also covers the enterprise mobility management features of Microsoft Device Management (MDM) for Office 365 that aren't included in EMS. You can learn more about the products included in EMS in the first book in this series, *Enterprise Mobility Suite: Managing BYOD and Company-Owned Devices (https://aka.ms/EMSdevice/details)*.

Azure Active Directory Premium

Azure Active Directory (Azure AD) Premium is a Microsoft cloud-based service that provides comprehensive user identity and application access management capabilities. Built on the rich set of directory-service features of Azure AD that is included in all Microsoft Azure subscriptions, the Azure AD Premium subscription includes additional capabilities for enterprise-level identity management. One of the most popular features of Azure AD Premium is its integrated single sign-on (SSO) support for thousands of popular Software as a Service (SaaS) apps. This means that instead of users having to use multiple sets of user names and passwords to access apps such as Salesforce, Concur, or Workday, they can use a single user name and password for a consistent experience across every app and device.

In addition to the features in the Azure AD Free and Basic subscriptions, the Premium subscription includes the following:

- Self-service group management that users can use to create and manage customized user groups
- Advanced security reports and alerts based on machine-learning that organizations can use to monitor and protect access to cloud applications
- Multi-factor authentication (MFA) that supports configuring user verification steps in addition to a single user name/password authentication process
- Microsoft Identity Manager (MIM) support option you can use if you need to configure additional on-premises hybrid identity services
- Password reset with write back for user self-service password management with on-premises directory services
- Azure AD Connect Health to monitor on-premises identity infrastructure and synchronization services available through Azure AD Connect

Microsoft Intune

Microsoft Intune is another Microsoft cloud-based service that provides mobile device management (MDM), mobile application management (MAM), and Windows PC management capabilities. Supporting Android, iOS, and Windows-based devices, Microsoft Intune also can be used as a standalone cloud service or connected to an existing on-premises Microsoft System Center Configuration Manager 2012 R2 or later deployment. Additionally, Microsoft Intune provides the infrastructure support for enterprise mobility management features included with Office 365.

Microsoft Intune supports a comprehensive mix of MDM and MAM capabilities, including

- Simplified device enrollment for Android, iOS, and Windows devices
- Mobile device management through configuration and compliance policies
- Device access profiles for managing access to virtual private networks, wireless networks, email servers, and certificate-controlled resources

- Conditional access to Microsoft Exchange Server or Exchange Online–based email accounts
- Mobile application deployment, installation, and management
- Mobile device lock, remote PIN reset, complete device factory reset, or selective wipe of company data while leaving personal data intact

Azure Rights Management

Azure Rights Management (Azure RMS) is a cloud-based service that helps you protect your organization's sensitive information from unauthorized access and controls how this information is used or shared. Using encryption, identity, and authorization policies to secure files and email, Azure RMS applies policies and permissions directly to files and email messages, independent of where they are located. Permissions follow files and email messages inside or outside your organization, networks, file servers, and applications. This behavior enables users to access company data no matter what device they use or how the data is shared.

Microsoft Advanced Threat Analytics

Cyberattacks and Internet-based threats have grown more and more sophisticated and continue to increase in frequency and severity. Organizations realize now more than ever that they need to be proactive in their efforts to protect corporate data, user identities, employee and customer personal information, and their online reputation. Advanced Threat Analytics (ATA) identifies suspicious activities and abnormal behavior in on-premises networks, helps detect malicious attacks, and provides alerts for security risks. ATA is covered in more depth in Chapters 4 and 5.

Mobile Device Management for Office 365

The Office 365 business productivity suite is a group of cloud-based services and software subscriptions designed to increase productivity and lower licensing costs for organizations of all sizes. Office 365 and EMS are complementary suites of services and share many of the same architectural services. By sharing a common cloud-based infrastructure, both suites offer identity management provided by Azure AD, mobile device and application management capabilities provided by Microsoft Intune, and access and information protection enabled by Azure RMS. Microsoft ATA is an on-premises service and is included with EMS, but it isn't currently included with Office 365 subscriptions. Table 1-1 shows the relationships of these services.

Mobile Device Management for Office 365 (MDM for Office 365) is the group of mobility device management features included as a part of most Office 365 subscription plans. MDM for Office capabilities are enabled by Microsoft Intune and mobile device management features are tightly integrated with Office 365 services like Exchange Online and SharePoint Online. Instead of using the Microsoft Intune management portal, MDM for Office 365 management is built into the Office 365 admin console. Details about MDM for Office 365 will be covered in more depth in Chapter 6, "Introducing Mobile Device Management for Office 365," and Chapter 7, "Implementing Mobile Device Management for Office 365."

TABLE 1-1 Enterprise Mobility Suite and Office 365 products and services

	Enterprise Mobility Suite	Office 365
Identity management	Azure AD Premium ■ Single sign-on for SaaS apps ■ Advanced multifactor authentication ■ Microsoft Identity Management (MIM)	Identity management enabled by Azure AD ■ Basic single sign-on for Office 365 ■ Basic multifactor authentication for Office 365
Mobile device and app management	Microsoft Intune ■ MDM and MAM support ■ Advanced device and app policies ■ System Center integration	MDM for Office 365 enabled by Microsoft Intune ■ Basic device settings management ■ Selective wipe/device reset ■ Built into Office 365 Management Console
Access and data protection	Azure RMS ■ Protection for content in Office apps (on-premises or Office 365) and Windows Server files ■ Email notifications for shared documents	RMS protection enabled by Azure RMS ■ Protection for content in Office apps (on-premises or Office 365) ■ Access to RMS Software Development Kit (SDK)
Threat protection	Advanced Threat Analytics ■ Detects abnormal user behavior ■ Detects malicious attacks ■ Identifies known risks	Advanced Threat Analytics ■ Detects abnormal user behavior ■ Detects malicious attacks ■ Identifies known risks

Selecting the best solution for your organization

Determining the enterprise mobility management solution that best fits the needs of your organization can be difficult. As you've seen from the brief overview of the capabilities of the Enterprise Mobility Suite and MDM for Office 365, differences exist between the features and capabilities for each suite (and component services). Choosing one over the other before thoroughly understanding the differences and matching each to the specific needs of your organization will likely result in wasted time, wasted money, and user dissatisfaction and frustration.

In this section, you'll cover key enterprise mobility management planning and design considerations you'll need to define to choose the best Microsoft solution for your organization. Additionally, you'll compare the features and capabilities of the Enterprise Mobility Suite, Microsoft Intune (standalone deployment), and MDM for Office 365.

Planning and designing a solution

The first step in determining what enterprise mobility management solution best meets the needs of your organization is defining your requirements. These requirements aren't just a list of mobility management capabilities you think your organization needs—they must meet the actual business and productivity needs of your organization and users. You'll need to review the functional and service capabilities of each solution to answer questions in the following areas:

- Business needs, including device ownership, device platform, application, and user requirements
- Mobility management location needs, including geographic network requirements
- Mobile device management life-cycle requirements, including device enrollment, configuration, security, management, and monitoring
- Software as a Service (SaaS) connectivity requirements

Defining your business needs

To get started, you must understand your current and future business needs and how they fit with your organization's business strategy. If you don't take a long-term approach with your mobility management planning, chances are that your solution won't be scalable as your organization changes and grows. Although each organization will have different business requirements, a good place to start is to leverage best practices from other organizations in your industry. Because more and more organizations are embracing mobility management solutions with each passing day, it's likely that mobility management resources are available to help you with this planning. If your organization is regulated by governmental agencies or need to meet industry-specific compliance standards, you should review the applicable standards for your organization for any mobility management requirements or guidance.

Next, it's time to match the business requirements you've defined to the specifics of mobile device management:

- **Device ownership**   Who will own the mobile device? The employee, the company, or a mix of both options?
- **Device platform**   Which mobile device operating systems need to be supported? Just one or a mix of several?
- **Applications**   Which mobile applications or SaaS apps need to be supported? Are the applications supported on the required device platforms? How will the applications be deployed?
- **Users**   Will different groups of users have different mobility needs? Will users need mobile access to the same resources accessible from on-premises workstations?
- **Compliance**   How will compliance requirements affect mobile applications? Are management policies in place for mobile devices? Does your organization already have a BYOD policy in place, or will you need to create one?

Defining your location needs

Location can affect the administrative model of your mobility management solution. Some solutions support only a cloud-based service model, while others support a hybrid cloud/on-premises administrative model, where cloud-based services are connected to on-premises solutions. Depending on your network infrastructure and the geographic location of your company offices, having the flexibility to connect to existing device management solutions and use a central point of administration can significantly reduce costs and administrative overhead.

Modern mobile devices almost always include Global Positioning System (GPS) features by default. These features enable mobile applications to leverage geolocation capabilities. Some organizations might have business scenarios in which disabling location services on mobile devices is a requirement. For example, a company might have employees working in areas where applications that use location services cannot be used because of the sensitive nature of the work. IT departments would need to disable location services on devices that have access to these areas.

Answer the following questions about your location requirements:

- **Administrative model** Which administrative model best meets your current and future infrastructure needs, centralized or distributed?
- **Location services** Does your organization need the ability to disable location services on mobile devices?

Defining your mobile device life-cycle requirements

Managing mobile devices, both company-owned and user-owned devices, encompasses several important life-cycle management decisions. You need to define how your organization will manage devices in the each of the areas shown in Figure 1-3, making sure that each aligns with your overall MDM strategy, business needs, and other network management and support policies.

FIGURE 1-3 Mobile device management life-cycle stages

ENROLL

Mobile device management starts with enrollment, and it must be simple, easy, and reliable. If device enrollment is complicated, difficult, or unreliable, users will be resistant to following the process or slow to enroll their mobile devices for management. Typically, devices are registered with a mobility management solution either by a user self-enrollment or an administrator-managed bulk-enrollment process.

In the self-enrollment process, users enroll devices by accessing an enrollment or management portal. This is a manual process, and organizations need to provide users with clear enrollment guidance to avoid creating additional support cases. In most cases, IT will require users to enroll their devices if they want to access corporate resources from their mobile device. For example, most users want to immediately configure access to their work email account from their mobile device, and policies can be configured to automatically provision user devices to access corporate email when they enroll the device.

Answer the following questions about your device-enrollment needs:

- Will mobile devices be enrolled by administrators, by users, or by both?
- Does your organization need to bulk-enroll devices?
- How many devices will each user typically use and need to enroll?
- What are the connectivity requirements for users to self-enroll devices?
- What are the enrollment requirements for each device operating system your organization needs to support?
- Do you require special policies for device-enrollment failures?
- Will IT and users both need to unenroll devices?
- If a device is selectively wiped, should it automatically be unenrolled from management?

CONFIGURE

The configuration and compliance policies in the mobility management solution must align with the business requirements for your organization. Typically, a mobile device is automatically assigned these policies and permissions when the device is enrolled, and administrators can associate these policies with groups of either devices or users.

Answer the following questions:

- Which internal and external applications and services will be deployed, managed, and accessed by mobile devices?
- What mobile device security and access configurations do you need to enforce?
- Do you need to deploy apps and agents automatically and manually?
- Do you need separate levels for device-management permissions for IT roles and positions?
- Will your organization require digital certificates to authenticate mobile devices to company resources?
- How will mobile devices connect to the Internet when connected to the company network?

SECURE

Although usage of mobile devices can increase employee productivity, it also can increase security threats that you'll need to mitigate to protect your company's data and maintain user privacy. Defining your organization's data-protection requirements for mobile devices is an important planning step to address this concern. You should plan for mobile-device encryption (for both in-transit and at-rest data), data segregation, and device hardening. Each of these high-level areas build on other protection-related design considerations that need to be defined.

Each mobile-device operating system you plan to support can.also control and protect devices using different methods and different levels of granularity. For example, if one operating system has more options for hardening the device than another, you need to define a common set of hardening options to protect each type of device. These hardening options can include defining custom compliance policies for device passwords, sign-in attempts, and encryption settings.

Maintaining user privacy and properly classifying data stored on devices is equally important. Your organization might already have privacy standards and policies in place for workstation computers, and these should extend to mobile devices. This is especially important when conducting device hardware, software, and file inventories. A clearly defined, transparent privacy policy outlining what, when, and how data is collected from mobile devices will ensure that users are comfortable about what information is shared with the organization. This policy should also establish clear boundaries regarding what is considered company data and how it will be protected.

> **MORE INFO** For more information about privacy and compliance, visit the Microsoft Intune Trust Center at *http://www.microsoft.com/server-cloud/products/intune-trust-center/faq.aspx*.

Defining who and which devices will have access to company data will also need to align with your organization's standards and policies. This access is controlled by establishing authentication and authorization policies in the mobility management solution. To control access for resources, the solution must verify that users are who they claim to be (authentication) and determine whether they should have access to the resource (authorization). Once these steps are completed, the solution must validate both the level of access the user will have for the resource and that the device accessing the resource complies with company policies.

No matter how carefully these security principles are configured, you need to plan for potential security incidents. If your organization is just getting started with mobility management, make sure that any existing security incident-response policies and requirements apply to mobile devices and that the mobility management solution supports meeting these requirements. Especially in larger organizations, mobility management responsibilities might be assigned to a department or personnel not normally accustomed to responding to security incidents. It's a good idea to involve your organization's security team early in the mobility management planning and design process to prevent this from occurring.

When defining your security requirements, answer the following questions:

- How will data be protected on devices at rest and in transit?
- Will your organization need data encryption for devices and data within applications?
- Will you need the ability to erase company data from devices, while preserving personal data on devices?
- What level of device-hardening settings do you need?
- How will you communicate the organization's privacy policy to mobile-device users?
- Where will mobile-device data be stored? Only on the device or also in the cloud? How is privacy managed in these locations?
- Do you need to classify data on mobile devices? Does the classification travel with the data or apply only to data on the device?
- How will you authenticate users? Will you need multi-factor authentication features?
- Does your organization use an on-premises Public Key Infrastructure (PKI) to issue certificates? How will this apply to mobile devices?
- Will you need to control access to mobile apps? Does access need to have different levels of control?
- How will lost mobile-device incidents be handled? If the device is compromised, what policies will ensure that malicious activity doesn't spread to other devices or the larger network?
- How will you be notified of security incidents? Proactively or in real time?

MANAGE

As you've just seen, mobility management security integrates with virtually every facet of an organization's technology infrastructure. Managing mobile devices is a topic that's just as broad and comprehensive. Mobile device management typically involves several administrative and management areas, such as configuring devices, managing applications, configuring access to networks and resources, and monitoring and reporting. In most mobility management solutions, configuration policies are used to define general organizational settings for devices and compliance policies enforce requirements for resource access. Additionally, conditional access policies can define access to specific services, such as email or file-sharing resources.

To simplify and standardize enforcement of these policies, many mobile management solutions use profiles to push settings for networks and services to mobile devices. For example, by setting up and deploying email profiles, IT departments can automatically configure mobile devices with the appropriate email server connection information. This arrangement helps users connect to the correct email server without having to remember specific connection details. Profiles can typically be configured for virtual private network (VPN) and Wi-Fi network access and certificate management.

Answer the following questions when defining your MDM management requirements:

- Do you need specific policies applied to groups of users, groups of devices, or groups of device operating systems?
- Will you need to apply separate policies based on whether devices are company or user owned?
- Will you need customized policies for network access? Email access?
- Do policies need to be exported to third-party security devices?
- Do you need a customized company portal for users to install apps?
- How will policies be used to manage access to on-premises or cloud-based resources?

MONITOR

Capturing and monitoring event and status information from mobile devices is vital to ensuring that users and devices comply with your organization's policies and standards. This is especially important for organizations that must comply with government or industry-specific requirements and guidelines. Reporting also can assist with inventory management and provide detailed information about installed software, hardware capabilities, and licensing compliance. Remember the importance of user privacy discussed earlier, particularly for user-owned mobile devices. Your mobility management solution shouldn't monitor, capture, report, or share any personal activity or information without the consent of your users. You need to be able to answer the following monitoring questions:

- What kind of reports will you need for mobile devices?
- Will reports need to be shared or accessed remotely?
- Are there specific issues or problems you will need to identify?
- Do you need customized or on-demand reports?
- After a device is unenrolled, should legacy information be archived or maintained?

Defining your SaaS requirements

Understanding how your mobile management solution will integrate with current or future cloud services is vital as more organizations leverage the scalability and power of cloud-based computing. This has a large impact on managing user identity and directory services. Connecting and synchronizing your on-premises directories with a cloud service is the driving force to uniting users, mobile devices, mobile apps, and mobile device management. Additionally, configuring and managing connections to third-party SaaS apps can be difficult and time-consuming if those tasks are not handled correctly and the connections are not properly maintained.

When defining your SaaS requirements, answer the following questions:

- Are business-critical SaaS applications available?
- How will your existing on-premises user and device accounts connect?

- Do passwords need to be synchronized with Azure AD?
- Will you implement single sign-on for your organization?
- What existing SaaS platforms do you currently use? Do they support specific mobility management solutions and features?
- How is user and device authentication handled? How are identity-related threats and anomalies addressed?

> **MORE INFO** For more information about planning and designing a mobile device management solution, see the "Mobile Device Management Design Considerations Guide" at *http://aka.ms/mdmdcg*. This guide covers many of the areas in this section in greater depth. You can also view a Channel 9 TechNet Radio presentation on design considerations for MDM by Yuri Diogenes and Robert Mazzoli at *https://channel9.msdn.com/Shows/TechNet+Radio/TNR1610*.

Comparing Microsoft mobility management solutions

Now that you've defined the mobility management requirements that meet the needs of your organization, you're ready to compare the requirements and features of Microsoft's enterprise mobility management services. You'll cover the main features and capabilities of the Enterprise Mobility Suite, Microsoft Intune, and MDM for Office 365 side by side so that you can easily compare them. However, because these are all cloud-based services and continuously updated, make sure you verify the most current features and capabilities of these services when you're ready to deploy a service.

Prerequisites

Make sure your organization and infrastructure meet the requirements of each mobility management solution:

ENTERPRISE MOBILITY SUITE

The main requirements for EMS depend on the individual requirements for each of the component services. EMS-specific requirements focus only on activation and licensing. The basic steps to activate EMS are as follows:

1. Sign up for EMS.
2. Activate a licensing plan.
3. Activate access.
4. Assign user licenses.
5. Deploy Azure AD, Microsoft Intune, Azure RMS, and Advanced Threat Analytics.

MORE INFO For more information about deploying EMS, see the Enterprise Mobility Suite Activation Guide at *http://aka.ms/EMSGetStarted*.

MICROSOFT INTUNE

Microsoft Intune is a cloud-based service, and there isn't a requirement to have any on-premises network infrastructure. Microsoft Intune uses the public Internet to communicate directly with devices and cloud-based users. If you do have an on-premises network infrastructure, Microsoft Intune will use your network to communicate with on-premises devices in your subscription. Although you are not required to use a dedicated server, options are available that use on-premises infrastructure components like Microsoft Exchange and Windows Server Active Directory tools. The basic requirements are

- **Mobile device platforms** Versions of Android 4 and later, iOS 7.1 and later, Windows Phone 8 and later
- **Computer platforms** Windows Vista and later versions (excluding Home editions)
- **Network ports** TCP 80 and 443

MORE INFO For more in-depth Microsoft Intune requirements, see "What to know before setting up Microsoft Intune" at *https://technet.microsoft.com/library/dn646966.aspx*.

MDM FOR OFFICE 365

MDM for Office 365 is simply a set of mobility management capabilities and requires only an Office 365 subscription. MDM for Office 365 requires

- An Office 365 commercial subscription (Business, Enterprise, EDU, or Government plan)
- Android 4 and later, iOS 7.1 and later, and Windows Phone 8.1 and later mobile devices

Features and capabilities comparison

Because the mobility management features in EMS are provided by Microsoft Intune, you really just need to compare the features and capabilities of Microsoft Intune and MDM for Office 365. The other component services of EMS don't provide mobility management–specific capabilities, though they do support mobility management–related capabilities.

Because many organizations use the basic mobility management features offered by Exchange ActiveSync, its features are included in Table 1-2 for a fuller comparison of features.

TABLE 1-2 Comparison of mobility management features for Exchange ActiveSync, MDM for Office 365, and Microsoft Intune

Category	Feature	Exchange ActiveSync	MDM for Office 365	Microsoft Intune
Device configurations	Inventory mobile devices that access corporate applications	x	x	x
	Remote factory reset (full device wipe)	x	x	x
	Mobile device configuration settings (PIN length, PIN required, lock time, and similar)	x	x	x
	Self-service password reset	x	x	x
Basic mobile device and app management	Provides reporting on devices that do not meet IT policy		x	x
	Group-based policies and reporting (the ability to use groups for targeted device configuration)		x	x
	Root and jailbreak detection		x	x
	Remove Office 365 app data from mobile devices while leaving personal data and apps intact (selective wipe)		x	x
	Prevent access to corporate email and documents based upon device enrollment and compliance policies		x	x
Premium mobile device and app management	Self-service company portal for users to enroll their own devices and install corporate apps			x
	App deployment (Android, iOS, Windows Phone, Windows 10)			x
	Deploy certificates, VPN profiles (including app-specific profiles), email profiles, and Wi-Fi profiles			x
	Prevent the cut, copy, paste, and save as operations from being used on data from corporate apps to share the data for use with personal apps (mobile application management)			x
	Secure content viewing via managed browser, PDF viewer, Imager viewer, and AV player apps for Intune			x
	Remote device lock via self-service company portal and via admin console			x
PC management	Client PC management (for example, Windows 8.1, inventory, antimalware, patch, policies, and similar)			x
	PC software management			x

Enterprise mobility management scenario

To help you understand of how each enterprise mobility management product will be used as part of the overall solution, the following scenario will be used throughout this book. Each implementation chapter will reference the scenario and implement one or more of the listed requirements. At the end of this book, you'll have the solution fully implemented, and it will meet all the requirements of the fictitious company shown in Figure 1-4.

FIGURE 1-4 Fictitious company's logo that will be used in this book

Blue Yonder Airlines recently completed some acquisitions and is expanding its business to different regions of the country. As result of these acquisitions, the company added several new branch offices around the country. Many of these branch offices are small (fewer than 50 employees) and don't have dedicated IT personnel on site, with some mobile devices currently being managed by an MDM solution the company will migrate away from and move to Microsoft Intune. The Blue Yonder Airlines IT department needs an easy solution to manage devices and applications to enable employees to be productive while physically located at the branch office and while out visiting customers. These acquisitions have also brought new challenges for application management because the scope of what needs to be managed has expanded to include new line-of-business and publicly available applications. To make things more difficult, some of the remote offices and remote users are bound by noncompete agreements for the next year and there are legal questions about how to incorporate these offices into the existing IT infrastructure.

Access to on-premises resources will likely increase, and Blue Yonder Airlines also wants to enhance its on-premises and cloud-based data-protection capabilities. After a meeting between the CEO and the CSO, it was agreed that part of this investment must also include ways to detect abnormal user behavior and malicious attacks, and to automatically identity known security issues and risks. Blue Yonder Airlines currently uses the Microsoft Enterprise Mobility Suite for its existing offices and has the following infrastructure components deployed:

- Windows Server Active Directory running on-premises and Azure Active Directory Premium in the cloud, with hundreds of users authenticating daily.
- Microsoft Intune subscription with security policies configured to manage existing BYOD scenario. All employees have devices enrolled, and there isn't an on-premises device management system deployed.
- Azure Rights Management Services (RMS) is configured for data access and protection.
- Exchange Server 2013 with users accessing their mailbox via Outlook client, native email clients included on mobile devices (using Exchange ActiveSync), and Outlook Web App (OWA).

Blue Yonder Airlines' goals for this project are to enhance its current Enterprise Mobility Suite deployment to ensure that data is more protected, applications can be more closely managed, and the new branch offices are included in the overall enterprise mobility management infrastructure. Blue Yonder Airlines established the following requirements in order to consider this project successful:

- Require remote office employees to have managed access to company resources and services from their personal mobile and corporate work devices, including remote offices bound by restrictive noncompete agreements
- Enable IT to enforce security, encryption, email, and device policy settings for remote offices using MDM for Office 365
- Monitor on-premises resources, and identify abnormal behavior in the network
- Prevent attacks to exploit known vulnerabilities for the resources located on-premises
- Detect security issues and risks, and alert administrators to them.
- Reduce false-positive alerts to avoid unnecessary red flags and distraction from real issues.
- Make full use of the Microsoft Intune mobile application management and data-protection capabilities mentioned earlier, and use it beyond basic app deployment.
- Define and implement a mobile-application strategy to support mobile device application deployment to all employee devices, regardless of whether they are managed by Microsoft Intune or not.
- Monitor app usage, and ensure the company complies with licensing agreements as the company grows through acquisitions.

CHAPTER 2

Introducing mobile application management with Intune

- The basics of app management with Intune   **22**

- Protecting apps and data with Intune MAM policies   **36**

- Managing applications without managing devices   **42**

In addition to managing devices, Microsoft Intune can also be used to manage the many mobile apps in use in enterprise environments. In fact, one of the primary reasons many companies use Microsoft Intune is to manage the multitude of apps that are available today. Mobile devices are everywhere, and it's up to you to ensure that your users can be productive on the device of their choice while, at the same time, ensuring company data is secure. In other words, you need to manage not only the device, but also the apps your employees are using to access company data.

Literally millions of mobile apps are available right now through various mobile-device app stores, with billions of app downloads over the past few years. Many of your users will access one or more app stores for each kind of mobile device they use on a regular basis. Mobile device users trust the app stores as the source for acquiring their apps, and using those mobile apps is becoming a standard business practice for many organizations. Mobile applications continue to become more ubiquitous in enterprise environments, and the explosion in the popularity of mobile devices drives not only usage, but also increased user expectations for both functionality and ease of use. Your end users want to be able to use their devices for both personal and company business, and they want that usage to be seamless. So it's clear that mobile apps are not going away, but how does IT embrace this trend without risking the safety of company data?

You are probably already familiar with Intune's mobile device management (MDM) capabilities that are used to configure devices and ensure that they are compliant with company policies, but what about apps and data? That's where mobile application management (MAM) comes into the picture. Instead of trying to manage the entire device, Intune's MAM capabilities are leveraged to manage some of the apps and data being used or accessed by company users. Using MAM policies, you can be sure that your users' personal apps and data remain their business while company apps and data, your business, remains secure and managed in accordance with company policies.

The basics of app management with Intune

Supported mobile devices and Windows PCs are only half the equation for enterprise users using various devices to get their jobs done. The other half is obviously the apps that they need to be productive, and Microsoft Intune's MAM capabilities ensure the equation remains in balance. Although company employees will certainly install many apps directly from public app stores (unmanaged apps), Microsoft Intune also can be used to install apps on end-user devices either automatically or on demand (managed apps). In addition to the standard unmanaged and managed apps, there's also another category of app management called *MAM-protected apps*. MAM-protected apps are managed apps with an additional layer of protection provided by Intune MAM policies.

Before diving into the deep end of using Intune to deploy applications to supported mobile devices and PCs and then manage them, you first need to gain a basic understanding of the different components used and requirements you need to meet to perform simple app deployments. Here's the list of tasks you need to perform:

- Prepare Microsoft Intune for app management and deployment.
- Define a group hierarchy to help organize users and managed devices.
- Publish apps to Microsoft Intune cloud storage.
- Leverage the right installation types and deployment actions to deploy apps to user or device groups.
- Monitor the app deployment.

> **MORE INFO** For the complete list of supported operating systems, see "Mobile device management capabilities in Microsoft Intune" (*https://technet.microsoft.com/library/dn600287.aspx*) or "Windows PC management capabilities in Microsoft Intune" (*https://technet.microsoft.com/library/dn646975.aspx*), which detail systems supported in Microsoft Intune.

Set the mobile device management authority

The first thing that must be done before deploying apps with Intune is to set the mobile device management (MDM) authority to Microsoft Intune. You do this on the mobile device management page of the ADMIN workspace in the Intune administrator console.

By default, there is no MDM authority configured for the service. When you set the MDM authority to use Microsoft Intune, the service is automatically configured with permission to manage a set of devices, which enables device enrollment, app management, and other functions for the service.

> **IMPORTANT** Several options are available for setting the MDM authority to define whether you will manage devices with Microsoft Intune standalone (covered in this chapter), System Center Configuration Manager with Microsoft Intune integration (hybrid MDM), or MDM for Office 365 (which is covered in Chapter 6, "Introducing Mobile Device Management for Microsoft Office 365," and Chapter 7, "Implementing Mobile Device Management for Microsoft Office 365"). Choose wisely.

Create user and device groups

Groups are used with Intune to logically organize collections of users or devices (both computers and mobile devices). These groups are then used to target management policies, apps, updates, and other content.

User groups can contain users synchronized from on-premises Active Directory security groups, or you can manually create a user group based on membership rules you specify. To add a device to a group, the device must be enrolled in management or discovered by Exchange ActiveSync. You can create groups of devices or groups of users, but you cannot create a group containing both. Other than abiding by that limitation, you are free to create groups to help you organize users and devices in any way you choose. That can be by user geographic location, device type, business units, or whatever other organizing principle you want to use.

You can create, modify, or delete custom groups in the GROUPS workspace of the Intune administrator console, but by default the service provides nine built-in groups that you cannot modify:

- All Users
- Ungrouped Users
- All Devices
- All Computers
- All Mobile Devices
- All Direct Managed Devices
- All Exchange ActiveSync Managed Devices
- All Corporate-Owned Devices
- Ungrouped Devices

Each group you create must be a child of an existing group, and you cannot change this relationship after the group is created. You can create as many child groups as you like in your custom groups, but know that if you delete a parent group, all the associated child groups are deleted with it. In addition, the default scope for users or devices when creating a child group is the membership of parent groups. So child groups are generally smaller subsets of resources taken from the available parent group's user or device membership.

> **TIP** You can add users or devices to child groups that are not members of the parent group. However, if you do that, they are also added to the parent group automatically.

Groups can contain or exclude specific users or devices based on group membership rules you specify when you create or modify a group. These membership rules can be based on some common criteria, or you can directly add a user or device to a group, as explained here:

- *Criteria membership rules* include or exclude group members based on security group or manager information synchronized from on-premises Active Directory.
- *Direct membership rules* are used to manually include or exclude specific group members in addition to the group membership resulting from the criteria you previously set.

Getting apps to the cloud

Because Intune is a cloud service, you first need to get the apps you want to deploy and manage on devices into some kind of cloud storage. And, among other things, that's what the APPS workspace, Microsoft Intune Software Publisher, and Microsoft Intune cloud storage are all here to help you with.

The APPS workspace

The *APPS workspace* in the Microsoft Intune administrator console provides information about detected and managed apps. This workspace is divided into three pages, as shown in Figure 2-1:

- **Overview** You can use this page of the APPS workspace to look at app deployment status alerts, add apps you can then deploy to devices, or view reports detailing software inventory information from managed PCs.

> **TIP** Microsoft Intune inventories only software installed on Microsoft Windows PCs. The service does not perform a software inventory of apps installed on mobile devices.

- **Detected Software** Here you can view the software inventory information collected from managed Windows PCs. In addition to viewing inventoried application properties, you can add license agreements for detected software to help you track license usage for those apps on managed PCs.
- **Apps** You use the Apps page of the APPS workspace to do most of the app-management tasks using Intune. Here you add new apps to be deployed, manage existing app deployments, view or edit app properties, and finally, delete any apps you are no longer

interested in managing. This page is also where you can add any volume-licensing or other software agreements you might have to help track software usage in your enterprise.

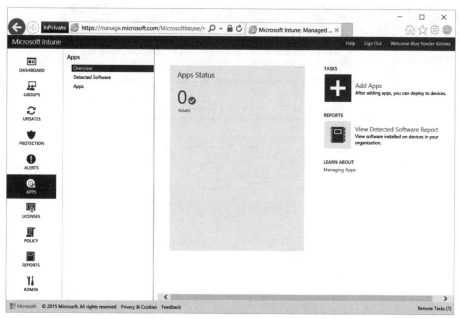

FIGURE 2-1 The APPS workspace in the Intune administrator console

Microsoft Intune Software Publisher

The *Microsoft Intune Software Publisher* starts when you add or modify apps from the Microsoft Intune administrator console. From the publisher, you select a software installer type that will either upload apps (programs for computers or apps for mobile devices) to be stored in Intune cloud storage or link to an online store or web application.

You can access the Microsoft Intune Software Publisher by clicking Add App on the Apps page of the APPS workspace. You'll get a security dialog box asking if you want to run the application from the Internet (at *manage.microsoft.com*). In this case, it's perfectly fine to do so. The software publisher downloads and opens automatically. You are prompted to log in with your Intune admin credentials, and then the software publisher opens and presents you with a page similar to the one shown in Figure 2-2.

When the software publisher opens,.the Software Description page is displayed. Here you can enter metadata information and details about the app that you want to publish to cloud storage. In addition to filling in the mandatory publisher, name, and description fields, you can also select a category from the following options: Other Apps, Books & Reference, Business, Collaboration & Social, Computer Management, Data Management, Development & Design, Photo or Media, or Productivity. On this page, you can also add an icon that displays for the app and decide whether or not to display the app as a featured app and highlight it in the company portal, as shown in Figure 2-3.

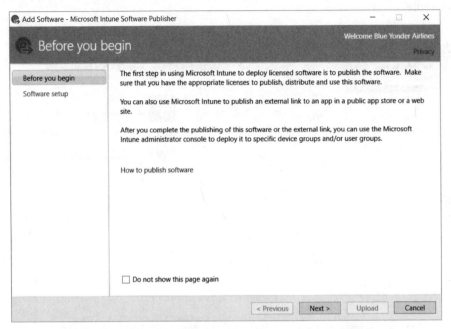

FIGURE 2-2 The Before You Begin page of the Microsoft Intune Software Publisher

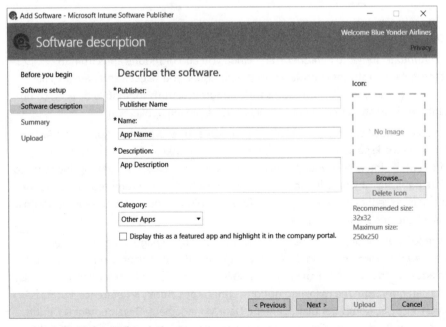

FIGURE 2-3 The Software Description page

Next, if there are any requirements associated with the app, such as a specific mobile device type or operating system, you can add those on the Requirements page of the Microsoft Intune Software Publisher. After you review the app details on the Summary page, select the upload option to upload the app to your cloud storage, where it is ready for you to deploy to managed devices.

Finally, if you want to modify an app after you upload it to cloud storage, select the app and then select the edit option from the toolbar above the list of available apps. This launches the Microsoft Intune Software Publisher, where you can make any necessary updates and then re-upload the app with your requested changes to your cloud storage.

Microsoft Intune cloud storage

All apps and updates you deploy are packaged and uploaded to Microsoft Intune cloud storage. How much storage space you start out with depends on the type of Microsoft Intune subscription you are using. If you are using a trial version, you automatically have 2 GB of storage space to use during your trial testing; after you purchase a subscription, you will be granted a full 20 GB of cloud storage.

> **TIP** All files you upload to Microsoft Intune cloud storage must be no larger than 2 GB, and your Internet connection speed must be at least 768 kilobits per second (Kbps).

To see how much cloud storage you are using, just look at the Storage Use page in the ADMIN workspace of the Intune administrator console. That page will tell you how much storage space you are using from the available storage space allocated to your service instance.

If you discover that you need additional cloud storage space after you purchase a subscription, a global or billing administrator can purchase the Microsoft Intune Extra Storage add-on, which gives you the ability to increase your available cloud storage in 1-GB increments. To purchase the Microsoft Intune Extra Storage add-on, just log on to the admin portal and add the add-on to your subscription under the Billing And Subscription Management page.

> **NOTE** If you purchased Intune or the Extra Storage add-on through an enterprise agreement, you will need to contact your account manager or Microsoft Partner for pricing information.

Software installation types

Intune gives you the capability to deploy several types of software to managed devices and PCs. You need to know which type of installation to select, as determined by both the app and the type of device you want to deploy the app to.

You can use the Microsoft Intune Software Publisher to prepare apps for deployment using the following software installation types: *software installer, external link,* or *managed iOS app from the app store*. (See Figure 2-4.) You can use these installation types to initially deploy apps to users and devices as well as to manage app updates for previously deployed apps when they are updated to newer versions.

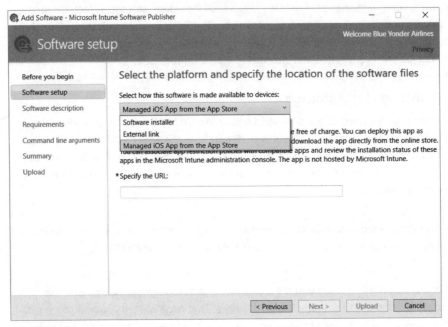

FIGURE 2-4 Making software available to devices on the Software Setup page

Software installer

The Intune software installer installation type option provides several options for installing applications on managed mobile devices, Windows PCs running the Intune computer client, or Windows 8.1 or later PCs being managed as devices. You use this installation type to upload a signed app package to cloud storage or install apps on managed mobile devices from an installation file without accessing an app store, in a process known as *side loading*.

As of the time of this writing, the following software installer types are supported for this method of app installation (as shown in Figure 2-5):

- **Windows Installer (*.exe,*.msi)** The installation files must support silent installation command-line options or installation with no user intervention. With that said, in most cases .msi files deployed by Intune do not require any command-line options to install successfully. You will also need to upload any additional files required for the .exe or .msi file to function properly.

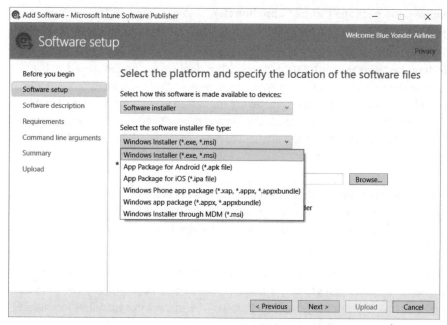

FIGURE 2-5 Selecting the software installer type on the Software Setup page

- **App Package for Android (*.apk)** Deploying apps to Android devices is straightforward once you set the MDM authority to Microsoft Intune and ensure you have a valid .apk file to deploy.

- **App Package for iOS (*.ipa)** For an .ipa package to be valid for deployment with Intune, it must be signed by Apple and the provisioning profile must not be expired. To create the app, your company must be registered for the iOS developer program and you will also need connectivity to the iOS provisioning and certification websites.

> **MORE INFO** You can learn more about the Apple developer program here:
> *https://developer.apple.com/programs/.*

- **Windows Phone app package (*.xap,*.appx,*.appxbundle)** These files must be signed by a Symantec enterprise mobile code-signing certificate before they can be deployed as side-loaded apps to Windows 8 and Windows 8.1 phones.

> **TIP** When side loading custom, line-of-business apps, Windows Phones trust only this one special type of certificate issued by Symantec and no other public or privately generated certificate. This is true for all mobile-device management solutions, not just Intune.

- **Windows app package (*appx,*.appxbundle)** The requirements for Windows app package deployment are the same as those for the Windows Phone app package deployments except you can deploy these to Windows RT and Windows 8.1 devices.

- **Windows Installer through MDM (*.msi)** This deployment option is used by the Intune service to install .msi files on Windows 10 devices. You can upload only a single .msi file for deployment, but app updates are supported when the .msi product code of each version is the same. The product code and product version information is also extracted from the .msi during deployment to make completing the required fields of the software publisher easier.

> **TIP** If you get an error saying that Microsoft Intune could not extract required information from the .msi file when using this installation type, try restarting the Intune console and software publisher by running your web browser as administrator.

Many software installer options present you with additional options to customize the deployment, such as requirements to begin an installation or command-line arguments to be used during the install. These settings are straightforward, and the software publisher explains those in great detail, so they won't be covered here.

External link

Probably the most commonly used installation type is the *external link*. Using this method, Intune deploys a software installation link to groups of users for an app hosted in a public app store. You can also use this installation type to provide a link to a web-based app that runs on a web browser. Users are then able to install the app from the company portal app as an optional installation if the app is targeted at their device type. In other words, you could technically deploy an external link app to all users, but only people using the device type associated with the app would actually see it in the company portal.

The process to find app URLs is relatively simple. Simply use a web browser to access the applicable public app store for the type of device you want to support, find the app, and copy the resulting URL. When deploying an external link installation type, you need to paste that URL in the appropriate text box as shown in Figure 2-6. After you specify the URL for the app, you just need to complete the app metadata, as shown in Figure 2-3 earlier, and sometimes configure a few other simple requirements, such as the supported operating-system type. You can then deploy this installation type as an available installation to groups of users managed in the Intune administrator console.

You can find an external link to use for this deployment type from one of these three app stores:

- Apple App Store for iOS devices: *https://itunes.apple.com/genre/ios/id36?mt=8*

- Google Play Store for Android devices: *https://play.google.com/store/apps/*
- Windows Phone Store for Windows devices: *https://www.microsoft.com/store/apps/ windows-phone*

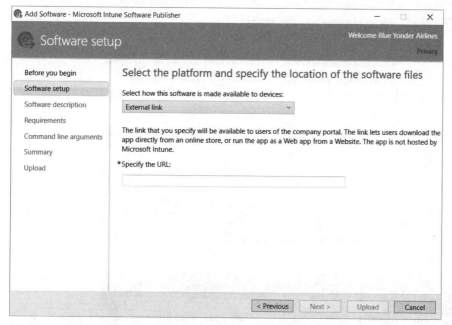

FIGURE 2-6 Selecting the External Link option on the Software Setup page

Managed iOS app from the Apple App Store

Similar to external links, managed iOS apps are deployed using links to free iOS apps from the Apple App Store, as shown in Figure 2-7. There are several differences, but basically what makes these iOS app deployments different from external links is that these apps can be managed much more stringently and deployed as required apps to managed iOS devices or groups of users.

In addition, when you deploy a managed iOS app you can associate mobile-application management policies, VPN profiles, or mobile-app configuration policies. Finally, you can review the deployment status in the administrator console just like you can with the other installation types.

When a managed iOS app is about to be installed on an iOS device (not in supervised mode), the device user will be alerted that the app will be installed as a managed iOS app. When the user clicks Install to begin app installation, she will be prompted to provide her Apple ID account password to download the app and install it from the Apple App Store, as shown in Figure 2-8.

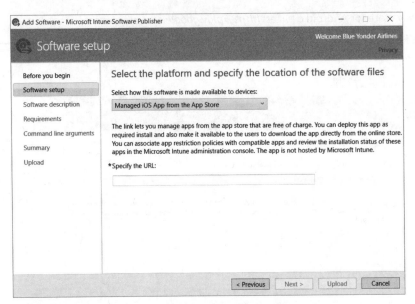

FIGURE 2-7 Selecting the Managed iOS App from the App Store

FIGURE 2-8 User prompts that appear when a managed iOS app is installed

If the app you are deploying as a managed iOS app is already installed when an installation begins, one of two things will happen:

- If the device operating-system version is less than iOS 9, you need to first uninstall the unmanaged app to allow the managed app installation to continue.

- If the device operating-system version is iOS 9 or later, you get a message asking if you would like to allow the currently installed app be managed, as shown in Figure 2-9. If you accept management of the app, the app and all data associated with it will be under Intune policy management and will be removed if the management profile is removed. In other words, if you unenroll your device, both the app and data will be selectively removed as well.

FIGURE 2-9 Message telling you when Intune takes over management of an app on iOS 9+

Understanding app deployment actions

When you deploy apps to groups of users or devices, the app becomes either available for users to install from the company portal, automatically installed, or even uninstalled. So, to do this right, you need to create groups that target specific users or devices you want to receive the app, and you need to choose the appropriate app approval and deadline options.

Approval options

When you are managing app deployments, you can choose from one of the following app-installation approval options, as shown in Figure 2-10:

- **Required Install**   The app will be automatically installed on devices with minimal to no user intervention required, and the app will not appear as an available app in the company portal. (End users will be prompted to accept the app before it is installed when using Android devices or iOS devices not in supervised mode.)

- **Not Applicable**   The app will not be automatically installed or made available to devices through the company portal.

- **Available Install**   The app will be made available for users to install on demand from the company portal.

- **Uninstall**   The app will be uninstalled on devices when an uninstall action is supported by the app.

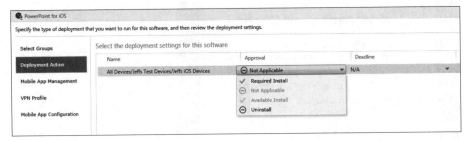

FIGURE 2-10 The available options for configuring app-deployment approval settings

Deadline options

When you are managing app deployments, you can choose from one of the following app-installation deadline options, as shown in Figure 2-11:

- **None**   Deploys the app based on the agent policy settings.

- **As Soon As Possible**   The Intune service will install the app on supported devices present in the groups targeted for app deployment during the next scheduled device policy synchronization activity. When you target a device for app deployment, the Intune service will immediately begin trying to contact the device to initiate the app installation.

- **One Week**   This option will deploy the app one calendar week from the current day.

- **Two Weeks**   This option will deploy the app two calendar weeks from the current day.

- **One Month**   This option will deploy the app one calendar month from the current day.

- **Custom**   This option lets you set a specific date and time for the app to deploy.

MORE INFO For more information about device synchronization information, see "Use policies to manage computers and mobile devices with Microsoft Intune" at *https://technet. microsoft.com/library/dn743712.aspx.*

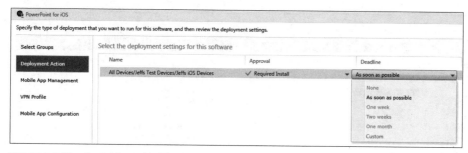

FIGURE 2-11 The Available options for setting an app-deployment deadline

Monitoring app deployments

Of course, once you deploy an app, you'll need to monitor how the deployment is going. You can easily do that from the Apps workspace of the Intune administrator console. If something has gone wrong, you will be able to see an alert clearly on the Overview page of the workspace, as shown in Figure 2-12.

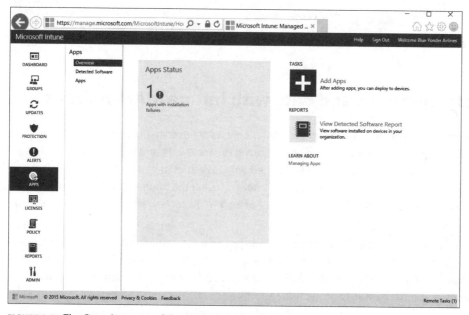

FIGURE 2-12 The Overview page of the APPS workspace

Selecting an app status alert takes you directly to the app properties. You can view how many users or devices have the app available or installed and also how many users or devices have experienced application failures. If you want to review other deployment status information, you can easily adjust the filter to review the app-deployment status information for both required desktop and modern applications, as shown in Figure 2-13.

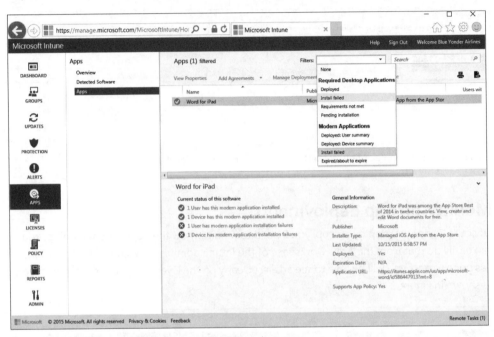

FIGURE 2-13 Available app filter options in the APPS workspace

Protecting apps and data with Intune MAM policies

Intune mobile application management (MAM) policies are used to provide an additional layer of data protection for managed apps running on Android (4 and later) or iOS (7 and later) managed devices. These policies can be applied only to restrict behaviors for apps that leverage the Intune App SDK or that have been "wrapped" using the Intune App Wrapping Tool for iOS or Android. Apps with this additional layer of built-in protection are called *policy managed apps* or *MAM-protected apps*.

When you apply policies to MAM-protected apps, you can configure settings that prevent data leakage, such as blocking standard copy and paste functions, preventing movement of data from a MAM-managed app to an unmanaged app, preventing backups to unmanaged locations, and other data-loss-protection settings.

There are several managed apps for both iOS and Android available in public app stores. Most notably, the Microsoft Office mobile apps—including Word, Excel, PowerPoint, OneNote,

OneDrive, and Outlook—come with included app-management capabilities enabled by Intune App SDK integration that you can easily leverage with Intune MAM policies.

> **MORE INFO** You can find a list of all Microsoft apps that you can use with Intune MAM policies on TechNet here: *https://technet.microsoft.com/library/dn708489.aspx*.

An added benefit included with many of these policy managed apps is that they support *multi-identity*. This means that Intune applies management settings only to company accounts or app data. For example, if a user configures his personal email account using the Intune managed Outlook email app, Intune will apply management settings only to the corporate account and ignore the personal email profile. Additionally, if the device is selectively wiped by an Intune administrator to remove all company data, the personal app data will remain present along with the now unmanaged app installation.

Creating MAM policies to protect company apps and data

Intune MAM policies are used to modify the functionality of apps by allowing you to control common app functionality like the copy, cut, and paste operations for managed apps. You can also use MAM policies to restrict device web-browsing behaviors when using a managed browser app. MAM policies are created in the Intune administrator console by navigating to the Configuration Policies page of the POLICY workspace and then selecting the Add option to create a new policy. Expand Software in the resulting new policy dialog box, select the type of policy you need, and then click Create Policy, as shown in Figure 2-14.

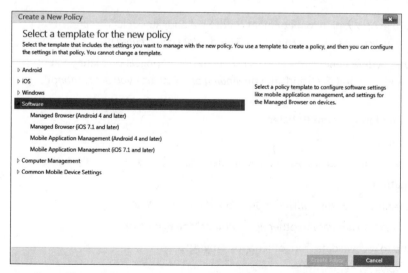

FIGURE 2-14 The available options to create software configuration policies

Unlike device-management configuration or compliance policies that are applied by deploying them to groups, Intune's MAM policies are instead associated with policy-managed apps during the deployment process. You can easily add a policy managed app to the Intune administrator console using the Intune Software Publisher, but you won't be able to deploy it without first associating a MAM policy with it. In fact, if you try, you'll receive an error reminding you of the fact that the app you are trying to deploy must be associated with a mobile-app management policy.

Create a mobile-application management policy

Mobile-application management policies are easy to create in the POLICY workspace of the Intune administrator console. You can create as many mobile-application management policies as you need in order to customize app restrictions on all of the different MAM-protected apps you deploy to managed Android and iOS devices.

When creating a mobile-application management policy for either Android or iOS, you have the choice of starting with a preconfigured policy using recommended settings or creating a custom policy from a completely new policy template. Before you can define a policy, you should have a deep understanding of your data-loss-protection needs to decide which policy settings to enable from the many available to you and also what values you want to provide for each setting.

The following are the available mobile-application management policy settings with their default values if you choose to start with recommended settings when creating a new policy (for creating either Android or iOS mobile-application management policies):

General Settings

- **Name**   Mobile Application Management (Android 4 and later *or* iOS 7.0 and later *depending on which one you are creating*) created *<date you created it>*.

- **Description**   A default policy created using the "Mobile Application Management (Android 4 and later *or* iOS 7.0 and later *depending on which one you are creating*)" template.

Mobile Application Management Policies

- **App Web Content**

  - Restrict web content to display in the Managed browser: **Yes**

- **Data Relocation**

  - Prevent (Android backups *or* iTunes and iCloud backups): **Yes**

  - Allow app to transfer data to other apps: **Policy Managed Apps**

  - Allow app to receive data from other apps: **Any App**

  - Prevent "Save As": **Yes**

  - Restrict cut, copy, and paste with other apps: **Policy Managed Apps with Paste In**

- **Access**
  - Require PIN for access: **Yes**
    - o Number of attempts before PIN reset: **5**
  - Require corporate credentials for access: **No**
  - Block managed apps from running on jailbroken or rooted devices: **Yes** *(iOS MAM policies only)*
  - Require device compliance with corporate policy for access: **Yes**
  - Recheck the access requirements after (minutes):
    - o Timeout: **30**
    - o Offline grace period: **720**
- **Additional Policies**
  - Encrypt app data: **Yes**
  - Block screen capture: **Yes** *(Android MAM policies only)*

Create a managed browser policy

The other mobile-application management software-configuration policy option available to you is the managed-browser policy for Android or iOS devices. This MAM policy sets and configures settings for the Intune managed-browser app for Android or iOS that you can find on the Google Play or Apple App Store, respectively. When you deploy the managed browser app, you simply associate a managed-browser policy with the deployment to ensure that device browsing activity remains compliant with company policies to either allow or block access to particular website addresses.

> **TIP** Because the managed browser is a MAM-protected app itself, you will still need to associate a mobile-app management policy for the managed-browser app when it is deployed and it will be subject to the same data-loss-protection settings you enable for the policy.

Of course, before you can deploy a managed browser using the software publisher, and then associate a managed-browser policy during its subsequent deployment, you will need to create the policy itself. The initial steps of creating a managed-browser policy are the same as those for creating a mobile-application management policy in the Intune administrator console. It's a good idea to know the URLs that you do, or do not, want managed browsers to view and which MAM-protected apps will be affected before creating these policies. Once you have the list of URLs you want to use, you once again have the choice of starting with a preconfigured policy using recommended settings or creating a custom policy from a completely new policy template.

The following are the available managed-browser policy settings with their default values if you choose to start with recommended settings when creating a new policy (for creating either Android or iOS managed-browser policies):

General

- **Name**   Managed Browser (*Android 4 and later or iOS 7.0 and later depending on which one you are creating*) created *<date you created it>*.

- **Description**   A default policy created using the "Managed Browser (Android 4 and later *or* iOS 7.0 and later *depending on which one you are creating*)" template.

Secure Web Viewer

- Enable an allow list or block list to restrict the URLs the Managed Browser can open (iOS 7.0, Android 4.0, and later): **Block the managed browser from opening the URLS listed below (block list)**

In addition to the default *block list* option that stops certain websites from being accessed on devices using the managed-browser app, you can also configure an *allow list*, which will block all URLs from being accessed by managed browsers except for those authorized by the policy, as shown in Figure 2-15. Note that you cannot enter both allow and block URLs in a single managed-browser policy.

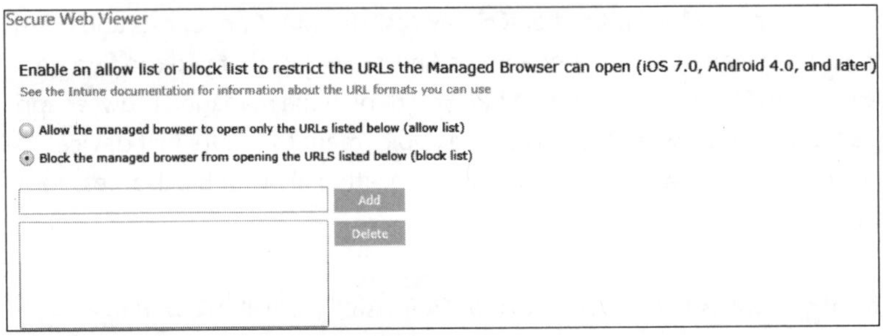

FIGURE 2-15 Secure web-browser policy settings

When adding URLs to the managed-browser policy, you need to be sure that you are adding the *precise* URL to either allow or block. Managed-browser policies will apply to attempts to browse only to *exactly* what you enter into these policy settings. If you want to allow or block an entire domain name space, be sure to use the * wildcard symbol in your allow or block list. For example, to allow or block all pages or subdomains for blueyonderairlines.com URLs, you enter **http://*.blueyonderairlines.com/*** for TCP port 80 web traffic and **https://*.blueyonderairlines.com/*** for TCP port 443 traffic. If you need to access URLs on nondefault http or https ports, you need to add that information into your managed-browser URL list as well. Wildcards are not supported for port information.

Creating a MAM-protected app of your own

Remember that the two ways an app can meet the requirements to become a MAM-protected app is either for them to be SDK managed apps like the Microsoft Office apps for iPad and _Android (that have the management capabilities built in to them through Intune SDK integration) or by *wrapping* an app providing that functionality on top of the finished app code.

Creating new mobile applications from scratch that incorporate the Intune App SDK can quickly become both time consuming and costly for most businesses. Instead, you can enable similar functionality for internal, line-of-business apps that your company develops by creating a wrapped app. A *wrapped app* is simply an in-house, line-of-business app that has been repackaged to include the Intune App SDK. Once an app is wrapped, you can change its default behaviors by using Intune MAM policies just like the apps that were originally created with Intune SDK integration.

To create a wrapped app, you use either the *Microsoft Intune App Wrapping Tool for Android* or the *Microsoft Intune App Wrapping Tool for iOS*. Using these tools enables you to create your own policy-managed apps and then publish them to Intune cloud storage for deployment. However, each of these tools has specific requirements that you will need to meet before you start wrapping your own apps.

Intune App Wrapping Tool for iOS

Companies that are developing their own in-house iOS apps can take advantage of the Intune App Wrapping Tool for iOS to enable advanced data-leakage-protection management functionality—without modifying the underlying iOS app itself. The tool is really just a Mac OS command-line utility that creates a "wrapper" around an existing iOS application. After the iOS app is wrapped, it will become a policy-managed app capable of consuming Intune's MAM policies. It's probably easiest to run this tool on the same Mac computer being used to develop iOS apps.

There are a few things that the wrapping tool does *not* do that you should be aware of. The tool cannot be used to process encrypted apps, unsigned apps, or apps with extended file attributes. It also cannot be used to wrap an app obtained from the Apple App Store. The following are some additional requirements you will need to consider as well:

- The app-wrapping tool must be run on Mac OS X 10.8.5 (or later).
- You need to have an Apple Enterprise Developer Account.
- You need to have an Apple signing certificate and provisioning profile.
- Your in-house app should be in the form of an .ipa or .app file for iOS 7 or later (or you should recompile them in Xcode to target a later version of iOS).

> **TIP** You can download the Microsoft Intune App Wrapping Tool for iOS here: *http://www .microsoft.com/download/details.aspx?id=45218*. You can learn more about using the Intune App Wrapping Tool for iOS on TechNet here: *https://technet.microsoft.com/library/ dn878028.aspx*.

Intune App Wrapping Tool for Android

Companies that develop their own in-house Android apps can also use a command-line tool to prepare their apps for Intune MAM policy management. The Intune App Wrapping Tool for Android is a Windows command-line PowerShell application used to enable data-leakage protection by wrapping the original app without modifying the underlying Android app itself.

This tool also has a few limitations to be aware of. Most notably, the Android app must not be encrypted, must not have already been wrapped by the app-wrapping tool, and must be written for Android 4.0 or later devices. It also cannot be used to wrap an app downloaded from the Google Play store. Additional requirements to run the App Wrapping Tool for Android include

- The tool must be used on a Windows PC running Windows 7 or later.
- The Android app to be wrapped must be an .apk file written for Android 4.0 or later.
- If the app to be wrapped is using the Azure Active Directory Authentication Library (ADAL), you must complete additional steps.
- The latest version of the Java Runtime Environment must be installed, and the Java path variable must be set to *C:\ProgramData\Oracle\Java\javapath*.

> **TIP** You can download the Microsoft Intune App Wrapping Tool for Android here: *http://www.microsoft.com/download/details.aspx?id=47267*. You can learn more about using the Intune App Wrapping Tool for Android on TechNet here: *https://technet.microsoft.com/library/mt147413.aspx*.

Managing applications without managing devices

By now, you are probably familiar with using Microsoft Intune to manage mobile devices and PCs managed as devices using the Intune administrator console, but what if you are already using a different MDM solution? What if you don't really want to manage devices at all and are just trying to find a way to manage the applications in use in your organization and to help protect company data? What if your users are dragging their feet or outright opposed to enrolling their personal devices in company management policies? Well, now you don't have to manage a device to manage the apps used on it.

Managing Intune's mobile application management policies from the Azure web portal gives you the ability to manage apps and protect corporate data on both managed and unmanaged devices. Even though you are using Intune to create and manage application management policies, those policies can manage applications and data on devices that are being managed by *any* MDM technology. Managing Intune application management policies from the Azure portal is not a replacement for MDM. Instead, this capability is meant to be used alongside MDM solutions—including non-Intune MDM solutions—that keep devices protected while Intune applies a set of app restriction and access policies to ensure app-layer protections are in place.

Because this capability is independent of any mobile-device management solution, you can use it to protect your company's data without having to enroll and manage devices. By simply implementing app-level policies, you can restrict access to company resources and keep data within the purview of your IT department. This capability is helpful in situations like when a company employee's mobile phone is enrolled in an MDM solution, but the employee is also using other, unmanaged devices to access company data from remote locations. In this situation, application management policies continue to protect company data without requiring the user to enroll all of his personal devices in management.

You can set MAM policies from the Azure portal that configure data-protection settings on MAM-protected apps to prevent company data from being transferred off the device or to unprotected apps. This includes the ability to restrict copy, paste, or open-in, and saving company data to unprotected storage locations. Office mobile apps supporting multi-identity continue to work as previously described to maintain company data within the context of the end user's corporate credentials. Personal data is kept separate.

Before you can start using this capability, there are a few important points to consider:

■ An Azure Active Directory or Microsoft Intune subscription is required to access the mobile application management settings. This is because these settings are not managed from within the Intune administrator console. Instead, they are managed from the Microsoft Azure web portal.

■ To apply app-management policies to company users, they must be assigned an Intune service license in the account portal.

■ When protecting apps on unmanaged devices, the user must install apps from the appropriate device-platform app store. You cannot initiate app installations for the user using this method.

How it all works

What makes Intune's mobile application policy management so simple to set up is the cloud-first architectural approach. When you configure MAM policy, Intune works with Azure Active Directory (AAD) to place a token in your AAD tenant. From that point on, when a user logs in to her corporate account on a MAM-enabled app, she gets both an authentication token granting access to the app as well as this new token that wakes up the Intune SDK and fetches policy.

In this way, the dependence between app distribution and app protection is separated. No matter how the end user receives an app (from the app store, MDM deployment, and so on), this process just works.

Once a policy is in place on an app, corporate data is protected and isolated from personal data. This includes data-leakage vectors between corporate and personal apps as well as within a single app. Think Outlook connected to a personal and corporate account—the SDK draws a boundary between these lines and enforces protection across it.

In this way, MAM policies are flexible enough to meet your BYOD (unenrolled) and CYOD (MDM enrolled) data-protection needs with a simple cloud-first set up.

Joey Glocke, Program Manager
Enterprise Client Management, Microsoft

After logging in to the Azure portal, you will probably not see the Intune mobile application management settings tile by default. However, to make it easier to find in the future, you can pin the Mobile Application Management tile to the portal start page for your Azure subscription. The easiest way to do that is to browse the available Azure services until you can find and select Intune. After you select Intune, you should see the Mobile Application Management blade, as shown in Figure 2-16. Once you open the blade, you can then select either Policy or Users to view the users available for policy assignment and also any existing policies you have created. This is also where you will add or modify policies.

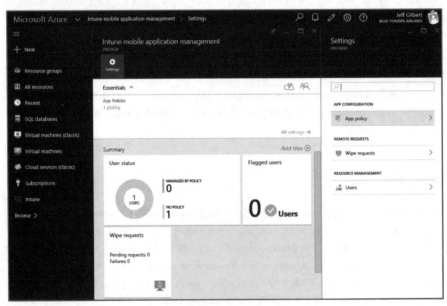

FIGURE 2-16 The Azure portal Startboard for the Intune mobile application management policy settings

At this point, you can create and deploy application management policies to Intune-licensed company users. Remember that these policies are associated with users and data, not with the device they use to access it. There are two categories of policy settings: data reloca-tion and access. (See Figure 2-17.) Data-relocation policies are applicable to data movement in and out of the apps, while the Access policies determine how the end user accesses the apps in a corporate context.

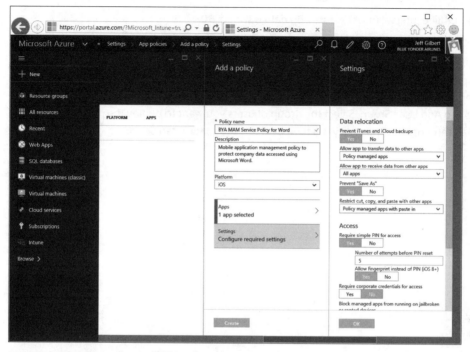

FIGURE 2-17 Creating an Intune MAM policy using Azure portal

The following are the available mobile-application-management policy settings, with their default values configured when creating a new policy for an iOS or Android app:

Data relocation

- Prevent (iTunes and iCloud *or* Android) backups: **Yes**
- Allow app to transfer data to other apps: **Policy managed apps**
- Allow app to receive data from other apps: **All apps**
- Prevent "Save As": **Yes**
- Restrict cut, copy, and paste with other apps: **Policy managed apps with paste in**

Access

- Require simple PIN for access: **Yes**
  - Number of attempts before PIN reset: **5**
  - Allow fingerprint instead of PIN (iOS 8+): **Yes** (*iOS only*)
- Require corporate credentials for access: **No**

- Block managed apps from running on jailbroken or rooted devices: **Yes**
- Recheck the access requirements after (minutes):
    - Timeout: **30**
    - Offline grace period: **720**
- Offline interval (days) before app data is wiped: **90**
- Block screen capture: **Yes** (*Android only*)

After you create a new mobile application policy, you still need to deploy it to users or groups of users before the policy will take effect. This is easy to do. Select the policy you want to deploy, select User Groups from the resulting blade that slides out, and then select the option to Add User Group. Select the group of users you want to deploy the policy to and then save all settings. The app policy will then be deployed to targeted users.

CHAPTER 3

Implementing mobile application management

I n Chapter 2, "Introducing mobile application management with Intune," you learned about mobile application management with Microsoft Intune and how it can be used to help keep company data secure while providing applications and access to your end users on the device of their choice. Armed with the prerequisite knowledge of the basics of Microsoft Intune's mobile application management (MAM) capabilities, an understanding of mobile application management policies, and even how to manage application data without managing the device, you should next begin to consider the additional implementation details required for your organization's specific needs.

In this chapter, you'll assume the persona of the senior enterprise administrator for Blue Yonder Airlines to plan, prepare, and implement the MAM strategy for the company as described at the end of Chapter 1, "Understanding Microsoft enterprise mobility solutions." You'll be responsible for making full use of Intune's mobile-application, data-protection, and reporting capabilities to support app and company data-access management for all Blue Yonder Airlines employees' mobile devices.

Scenario

As the senior enterprise administrator for Blue Yonder Airlines, you're responsible for planning, designing, and implementing the company's MAM solution. You need to ensure the solution requires as little administrative overhead or end-user disruption as possible while still enabling centralized reporting to monitor app management and software license usage. You also need to ensure you can protect company data on user devices— whether those devices are managed by Intune or not.

As part of modernizing Blue Yonder Airlines' application-management strategy, the company purchased Office 365 licenses to support new users brought in through acquisitions. You'll be responsible for deploying Microsoft Office mobile applications to those end-user devices. You'll need to install these apps on many devices, but some users already have installed the apps directly from the app store. You'll need to find a way to take over management of those installations while keeping your users' private data private and untouched by any MAM policies you implement. In addition to deploying the apps, you're responsible for managing application and data behavior on those end-user devices. Complicating matters further, you'll also need to manage app data behaviors on devices that are either unmanaged or managed by a different mobile device management service than Intune for new employees joining the company from the various acquisitions. Finally, your strategy must not only enable app and data management for devices in use by Blue Yonder Airlines employees, it must also enable you to monitor and respect all software-licensing agreements.

Implementation goals

With a basic understanding of mobile device management (MDM) and MAM capabilities available to you through Intune, you'll need to meet the following goals as you implement your MAM strategy for Blue Yonder Airlines:

- Plan for and roll out applications and policies in a way that does not disrupt the day-to-day work of Blue Yonder airlines employees.
- Deploy Microsoft Office mobile applications (Outlook, Word, PowerPoint, Excel, and OneDrive) on Blue Yonder Airlines iOS devices.
- Take over management ownership of any previously installed Office apps.
- Use mobile application management policies to manage app and data behavior for new deployments of Office apps on Intune-managed devices.
- Use mobile application management policies to manage app and data behavior for mobile devices not enrolled into management with Intune.
- Report on app installation status.

Solution diagram

To meet the Blue Yonder Airlines mobile application management implementation goals, you need to implement a solution addressing several considerations, as shown in Figure 3-1.

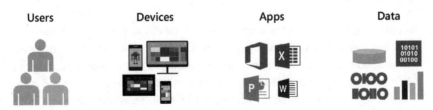

| Users | Devices | Apps | Data |

FIGURE 3-1 Elements of the Intune MAM solution

The elements are as follows:

- **Users**   Blue Yonder Airlines employees will be the end users for the mobile applications and application and data management policies that will be deployed. They must be made aware of the coming changes, and the app and policy deployments should not affect their productivity or personal data in any way.

- **Devices**   This solution design element represents both company-owned and employee-owned mobile devices that will be involved in the MAM strategy implementation.

- **Apps**   The Microsoft Office applications to be deployed on Blue Yonder Airlines' employee devices.

- **Data**   Company data on user devices must be safeguarded from data leakage outside of company-managed applications and storage. User personal data must also be effectively separated from company-managed data when they use the same apps for personal tasks that they use for company work.

Planning and designing the solution

Technically, you already know how easy it is to publish Microsoft Office mobile applications to the Intune service and deploy them to user mobile devices, but installing applications and restricting app and data behaviors can be a daunting task above and beyond the technical considerations. You need to ensure that you have your technical requirements covered and that you partner with the right teams and have communications in place to ensure the smooth rollout of apps and policies.

Before going into the Intune console, you need to plan and design a solution that fulfills your implementation goals, including goals for devices, apps, and data-access and app-data management.

Users

Of course, the first things you need to do before deploying any software are determine who needs access to it and ensure you are properly licensed to install it on users' mobile devices. Because the Office mobile apps for iOS come as a free download, your main concern is that there are enough Office 365 licenses available for your target users to actually use the apps you'll be installing. In our fictional scenario, Blue Yonder Airlines recently purchased enough Office 365 licenses to cover app installation and usage for all new employees. You will manage app deployments for these employees during this project, and you can be confident you'll be deploying and using the software legally.

You also know that some users are more tech savvy than others and more accepting of pilot-deployment conditions. You need to think about who will be involved in the early pilot deployments and who will be the most affected by the new data-usage restrictions.

Devices

Although many device types are in use in the Blue Yonder Airlines enterprise, you know that the primary devices you'll be targeting during this project are iOS devices. This is because the majority of mobile devices in use by the new acquisitions are iPads, which the Blue Yonder Airlines flight crews use daily. You will start rolling out the MAM strategy with these high-use devices and then move on to Android and Windows devices sometime in the near future.

Apps

Blue Yonder Airlines employees want a multitude of iOS apps and use many on a daily basis. However, for this project, you are tasked with ensuring that the Microsoft Office collaboration and productivity apps are installed and properly managed on iOS devices.

You need to plan to deploy and maintain Microsoft Outlook, Microsoft Word, Microsoft PowerPoint, and Microsoft Excel to ensure that users have the apps they need for day-to-day work. You also need to deploy and maintain Microsoft OneDrive to ensure company data is accessed and saved in a secure location. In addition, you need to deploy the Microsoft Intune Managed Browser app to ensure that web links stored in company data are opened in a managed and secure web browser. Your plan needs to account for employees using these apps for personal business too, and ensure that you can leverage multi-identity feature support available to you through these apps.

> **TIP** Some managed apps, including all Microsoft Office mobile apps, support multi-identity. Multi-identity allows you to use different accounts to separate work and personal data while using the same apps for both work and personal reasons. You can do this and also rest assured that Intune's MAM policies and data-wipe actions will affect only company data, leaving personal apps and data untouched and unmanaged.

Data-access strategy

After getting apps on user devices, you'll need to be confident in the data-access strategy that you want to implement to prevent company data leakage outside of managed apps and locations.

As part of your implementation, the data-access strategy must enforce policies to prevent company data from being copied or moved to personal apps. You also need to ensure that users cannot save data to unmanaged locations or view data on unmanaged applications. This means you need to manage where users can save data and force web links in company-managed data to be opened in a secure web browser. Finally, you need to ensure that the solution can remove company data from users' devices even if they use the same apps for company and personal business.

Unmanaged devices

The last planning challenge for you is to determine how you'll protect company data on unmanaged devices. Although many Intune-managed iOS devices will be part of this project, you also need to plan for how you'll support mobile application management on mobile devices that you do not directly manage with Intune.

You know that many devices being used by at least one of the acquired companies are managed by a different MDM technology. You also know many other employees simply will not enroll their personal devices into your management plan for whatever reason. For these devices, you must plan for implementing Intune MAM policies using the Azure portal to provide the required data-protection policies that will help keep Blue Yonder Airlines data secure and its employees productive on the devices of their choice.

Preparing apps and policies

Now that you understand the components involved and have completed the requisite preplanning to ensure your success, the time has come for you to prepare the apps and policies for the upcoming rollout to Blue Yonder Airlines employees.

In this phase of the project, you'll publish a managed iOS app using the Intune Software Publisher for each of the required Microsoft Office for iOS applications in preparation for deploying them. As you learned in Chapter 2, a managed iOS app from the app store requires a MAM policy to be applied to it if you want to manage it. So you'll need to create one of those as well. Finally, you'll need to create an Intune MAM policy in the Azure portal to manage company data in use on unmanaged devices in preparation for the rollout phase.

Publish the managed iOS apps

As you already saw in Chapter 2, publishing a managed iOS app is a straightforward process. The key points to remember about these apps is that they can be associated with mobile application management policies and deployed as required apps to managed iOS devices. If the app is already installed on an iOS device (iOS 9 or later), the user will be prompted to allow Intune to take over the management of the app. This makes the apps the perfect choice for this particular scenario.

Before you can deploy and manage the Microsoft Office apps for iOS, you first need to publish the apps to the Blue Yonder Airlines Intune service. Because you already identified the required apps, you just need to log in to Intune and publish the apps from within the APPS node of the Intune console using the Intune Software Publisher.

Find the app in the Apple App Store

The easiest way to find an iOS app is to search for it in the Apple App Store. Once you find it, you'll probably want to do two things. First, copy the web address to the app's location in the App Store because you'll need that to publish the app. Second, grab a copy of the app icon to use from this page as well. To keep things easy, copy and paste the default description into your app properties from here.

You can find the Microsoft Outlook app for iOS in the Apple App Store here: *https://itunes. apple.com/us/app/microsoft-outlook/id951937596?mt=8*. After you find Outlook, you need to find the rest of the Microsoft apps for iOS. You can find all the iOS apps Microsoft has published to the Apple App Store starting on this page: *https://itunes.apple.com/us/developer/ microsoft-corporation/id298856275*.

Publish the Microsoft Outlook for iOS app using the Intune Software Publisher

Now that you have the app location and probably an icon and description handy, you're ready to start the Intune Software Publisher and get to the business of actually publishing the apps to the Intune service.

To do this, log in to the Intune administrator console and click the Add App button at the top of the Apps page of the APPS workspace. The software publisher launches and then prompts you to log in with your Intune admin credentials. Once you're logged in, click Next on the Before You Begin page, as shown in Figure 3-2.

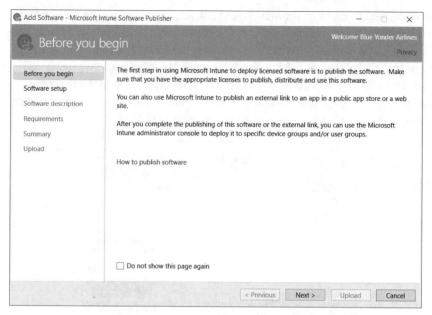

FIGURE 3-2 The Before You Begin page of the Microsoft Intune Software Publisher

On the Software Setup page, shown in Figure 3-3, select how the software you are publishing will be made available to managed devices. In this case, select Managed iOS App From The App Store from the drop-down list. After you do that, provide the web address for the app in the Apple App Store; just paste in the URL you copied earlier. Then click Next.

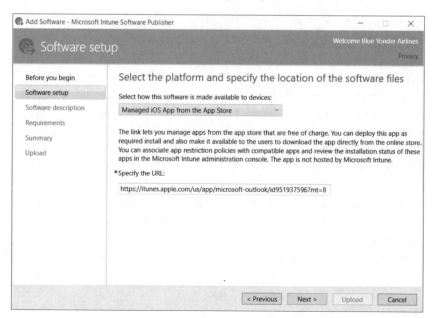

FIGURE 3-3 The Software Setup page of the Microsoft Intune Software Publisher

On the Software Description page, shown in Figure 3-4, you can fully describe the app and configure app settings such as the publisher name, app name, and a description of the app. You also can add an icon. These fields are easy to fill out because the information and graphics probably come straight from what you saw on the App Store page. This page is also where you can organize the app into a defined category (such as Collaboration & Social, as in this example), choose whether to display the app as a featured app, and highlight it in the company portal for your users to find on their own. After you enter the information, click Next.

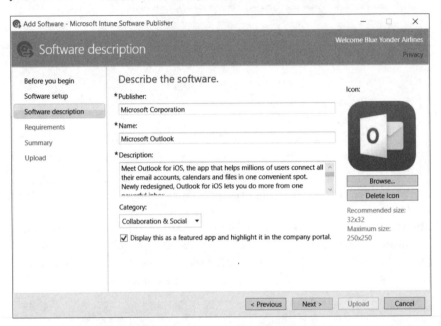

FIGURE 3-4 The Software Description page

Next, you have the opportunity to specify requirements that devices must meet for this app to be installed on them. In some cases, you don't want an iOS app going to all iOS devices because it might not work equally well on all device types. This is why the Requirements page of the software installer gives you a few options for device type. Specifically, it gives you these options: Any, iPad, and iPhone/iPod touch. Because the App Store description of the devices on which this app will work indicates both iPads and iPhones, leave the option set to Any, as shown in Figure 3-5.

Almost done! On the Summary page of the software publisher, you have the opportunity to review the information you provided up to this point and to go back to previous pages to correct any typos or make other changes that you might see, as shown in Figure 3-6. There is very little left to do now other than click Upload.

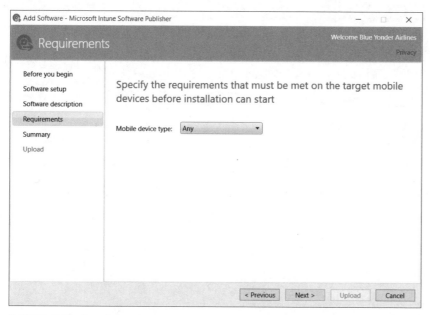

FIGURE 3-5 The Requirements page

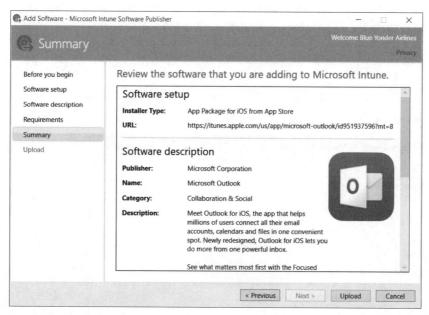

FIGURE 3-6 The Summary page

After you click Upload, the app data is published to the Intune service. You should see a green check mark signifying success, as shown in Figure 3-7.

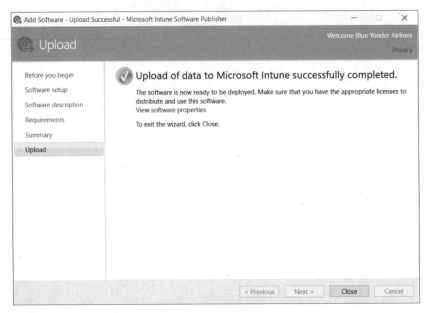

FIGURE 3-7 The Upload page

Click Close to exit the Intune Software Publisher Wizard. The app is now published to the Intune service and ready to be deployed to managed devices and users, but you might have to refresh the page a few times to see it. Once you've found it in your published apps list, it's easy to review the properties of the published app from the apps page of the APPS workspace, as shown in Figure 3-8.

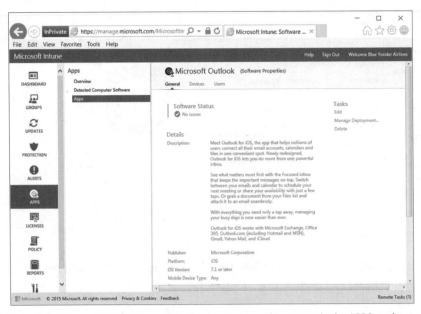

FIGURE 3-8 The published Microsoft Outlook app properties as seen in the APPS workspace

Publish the remaining apps

Now that Microsoft Outlook has been successfully published to the Intune service, you can begin publishing the other apps you've identified as necessary to meet the Blue Yonder Airlines MAM strategy implementation goals. Simply repeat the steps taken to publish the Outlook app for the following apps:

- Microsoft Intune Managed Browser (*https://itunes.apple.com/us/app/microsoft-intune-managed-browser/id943264951?mt=8*)

- Microsoft PowerPoint (*https://itunes.apple.com/us/app/microsoft-powerpoint/id586449534?mt=8*)

- Microsoft Word (*https://itunes.apple.com/us/app/microsoft-word/id586447913?mt=8*)

- Microsoft Excel (*https://itunes.apple.com/us/app/microsoft-excel/id586683407?mt=8*)

- Microsoft OneDrive (*https://itunes.apple.com/us/app/onedrive-cloud-storage-for/id477537958?mt=8*)

When you finish publishing all the required apps to the Intune service, the Apps page of the APPS workspace should display the full set of planned iOS apps ready to be deployed to Blue Yonder Airlines employees, as shown in Figure 3-9.

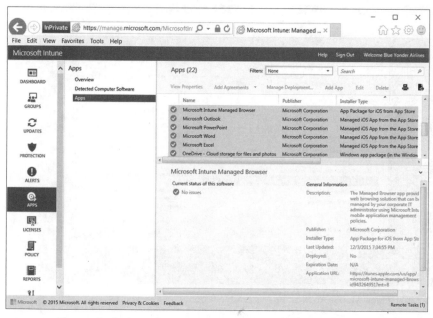

FIGURE 3-9 The required Microsoft apps published in the APPS workspace

Create a managed app policy to deploy with managed iOS apps

With the managed iOS apps from the App Store now published to Intune, the last step you need to complete to enable more granular control of the app and associated company data is to create the managed app policy that will be associated with the apps.

Although you'll be installing a handful of apps to your users, at a minimum, you need to have only one managed app policy to apply to each during the deployment of the app. Remember that the process flows like this: create an iOS managed app, create a managed-app policy, and then associate the managed-app policy with the managed app during the deployment phase.

Next, you need to log in to the Intune administrator console and navigate to the Configuration Policies page in the POLICY workspace. When you get there, click the Add option at the top of the screen to open the Create New Policy dialog box. Expand the Software category and select Mobile Application Management (iOS 7.1 And Later) option, as shown in Figure 3-10.

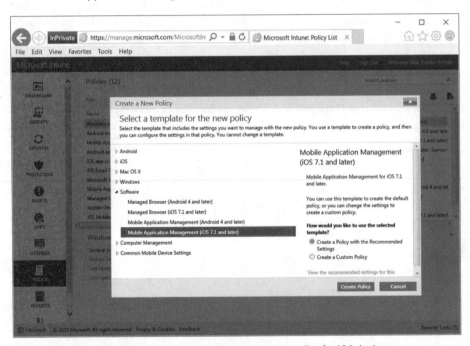

FIGURE 3-10 Creating a new Mobile Application Management policy for iOS devices

Next, leave the defaults selected to create a policy with the recommended settings and click Create Policy. Your request will be processed quickly and, if you scroll down to the bottom of your policies list, you should see a new policy called something like *Mobile Application Management (iOS 7.1 and later) create <today's date and time>*. Select the policy to highlight it. A notice appears in the console telling you that this policy cannot be deployed directly and that, instead, it must be associated with the software it will manage.

Double-click the new policy title to open it so that you can edit the default settings to match your deployment requirements, as shown in Figure 3-11.

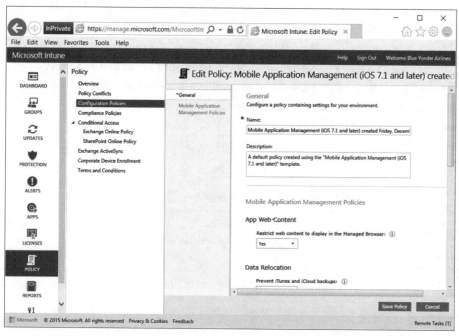

FIGURE 3-11 Editing a default mobile application management policy for iOS devices

The General policy settings section, shown in Figure 3-11, is where you provide the basic metadata for the new policy. You can give the policy a more interesting name and a better description than the autogenerated text telling you that the policy was generated using the policy template you started with. In this case, you can just shorten the title to something like *Mobile Application Management for iOS* and type a quick description like *Blue Yonder Airlines MAM policy for iOS apps*. The date is added to the title of a new policy by default, and that's probably something good to note in the description. With the date appearing in the description, you'll know when you previously made a change (and what it was). This information is particularly useful if that policy update causes problems for you later.

The remainder of the settings to configure for the MAM policy are contained in the Mobile Application Management Policies section and include App Web Content, Data Relocation, Access, and Additional Policies settings.

You use the App Web Content policy setting to configure what application behaviors occur when a user clicks a web link in company-managed data. The default setting for Restrict Web Content To Display In The Managed Browser is Yes. There's no need to change it, because part of your strategy is ensuring that web links stored in company data are opened in a managed and secure web browser.

The next policy section to configure is Data Relocation. This is the real heart of the policy, and luckily for you, the recommended settings are sufficient to meet your MAM strategy needs:

- Data from managed devices will be blocked from iTunes and iCloud backups.

- The managed app will transfer data only to other policy-managed apps.

- The managed app can receive data from any app but not the other way around, because you restrict cut, copy, and paste operations to only other managed apps.

- Users are also blocked from using the Save As function to save copies of company data to unmanaged locations.

When you finish configuring the App Web Content and Data Relocation sections, the policy settings should appear as shown in Figure 3-12.

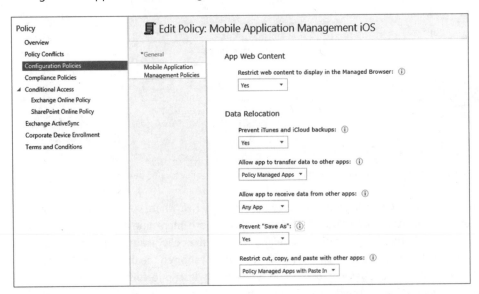

FIGURE 3-12 The App Web Content and Data Relocation settings of the Mobile Application Management policy for iOS

The next settings of the Mobile Application Management Policies section to configure relate to how the app is accessed by users. Here you decide to require a simple PIN to access the apps. This simple PIN is shared by all apps from the same publisher with the policy applied so that your users have to set this only once for all Office apps you'll apply the policy to. The other recommended settings are also fine to use because they support your MAM strategy goals. There's really no need for you to require that these apps ask users to provide corporate credentials and a PIN or that they recheck for access requirements more often than every 30 minutes.

The Additional Policies section contains the last setting to configure in this section. This is where you specify how app data is encrypted—as long as you have required a PIN. The recommended value, which you can easily accept, is to encrypt all data associated with managed apps, except for files the app currently has open, when the device is locked.

After you configure the last two settings sections, you should see something like Figure 3-13.

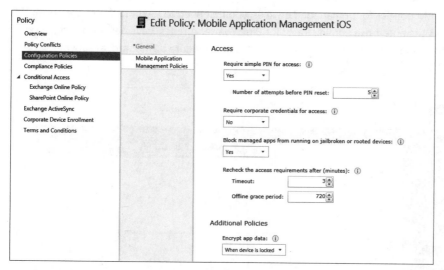

FIGURE 3-13 The Access and Additional Policies settings of the Mobile Application Management policy for iOS

After reviewing the policy settings one last time to be sure they accomplish your mobile application management policy objectives, click Save Policy at the bottom of the page. Your policy is ready to be associated with iOS apps during the deployment phase.

Create a managed app policy to deploy without deploying apps or managing devices

After publishing the managed iOS apps and creating a managed-app policy to deploy with them, you're ready to support app deployment for devices you manage with Intune. Remember that mobile device management ensures that devices are protected and that you can deploy managed apps to them, but mobile application management policies can also be used to ensure app-layer protections are in place whether the device is enrolled in management or not.

In some cases, you'll need to protect company data from being shared or stored inappropriately on your users' personal devices when they are not enrolled in management with Intune because they're either managed by other MDM technologies or they're just not managed at all. It's also common for users to use both managed and unmanaged devices to access company data. In either case, you need to be sure that a user reading her email at home on her smartphone keeps company data as secure as when using an Intune-managed device.

By configuring Intune MAM policies from within the Azure portal, you can provide that app-level data protection based simply on who users are instead of basing it on what device they choose to use. And you can achieve this level of protection without affecting end-user productivity or touching users' personal data in cases where apps support multi-identity.

Add Intune to the Azure portal services list

To create MAM policies for unmanaged devices, you first need to log in to the Azure portal. Because you, the enterprise administrator for Blue Yonder Airlines, have never done this, you might need to add the Microsoft Intune service tile to the Azure dashboard, as shown in Figure 3-14.

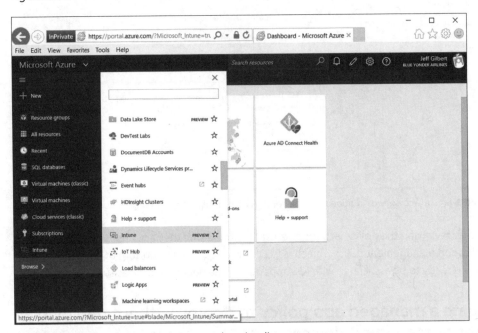

FIGURE 3-14 Adding Intune to the Azure portal services list

When you can see Intune in the Azure services list, you can select it to open the Intune mobile application management blade, which provides you with the tools necessary to create and manage MAM policies without requiring device enrollment. Of course, the first time you open the blade, you'll notice that there are no policies, as shown in Figure 3-15.

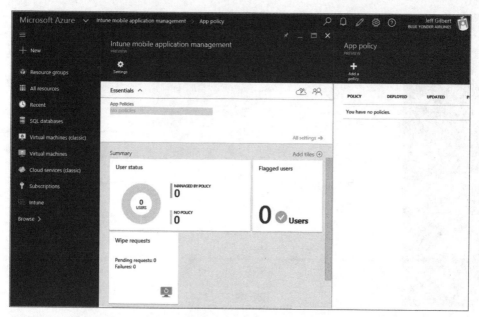

FIGURE 3-15 The default blade layout for Intune mobile application management in the Azure portal

Configure access roles

Before creating a policy, you should ensure you have the permissions required to do so and that you provided adequate permissions for others if necessary. By default, as a global administrator, you probably already have owner permissions, which allows you to manage everything, including access to all Blue Yonder Airlines resources. However, everyone needs a backup. In this case, you need to add Adam Barr, one of the Blue Yonder Airlines IT administrators, as a contributor to allow him to manage mobile application management policy tasks.

To add your backup administrator, click the Users icon at the top of the Intune mobile application management blade to display the Users blade, and then click Add to open the Add Access blade, as shown in Figure 3-16.

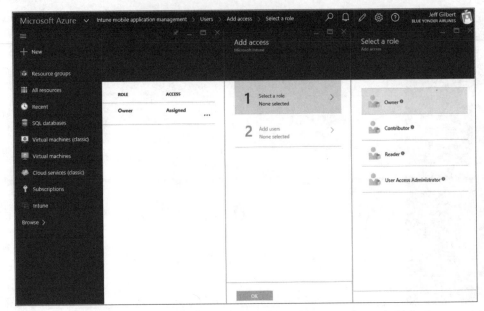

FIGURE 3-16 Preparing to add a user for the Intune mobile application management in the Azure portal

You know you want to add Adam as a contributor, so you click the Contributor role on the Select A Role blade and then use the search box to find his name in the user list. Select his name from the list, as shown in Figure 3-17.

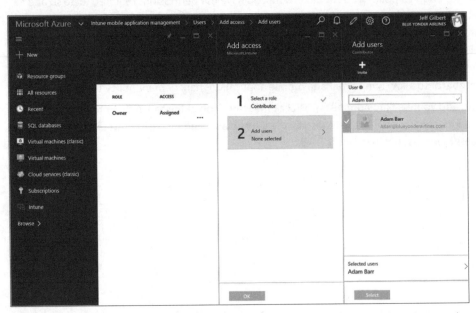

FIGURE 3-17 Preparing to add a user for Intune mobile application management in the Azure portal

Click Select on the Add Users blade, and then click OK on the Add Access blade. You should now see the new user account added as a contributor in the Users blade. With a backup in place, you're ready to create Intune MAM policies in the Azure portal.

> **TIP** This is a good time to ensure your entire MDM IT staff is included in some MAM management role.

Create a MAM policy to manage Office apps for iOS

With all the administrative details out of the way, the time has come to create the MAM policy in the Azure portal that you'll use to manage the way Office apps for iOS behave to protect Blue Yonder Airlines data from leaking between company and personal app usage. Your main goal here is to create a policy exactly like, or as similar as possible to, the managed application policy you created for Intune managed devices. You need to do this so that users are not confused by policy settings that vary depending on the device they choose to use and sometimes conflict with one another.

On the main Intune mobile application management blade, click Settings, then click App Policy on the Settings blade, and finally click Add A Policy on the App Policy blade. Once there, you should be ready to configure the general settings for the new MAM policy. If you keep things simple and provide the same information for this policy as you did for the managed-app policy (which you created previously in the Intune console to manage apps on managed iOS devices), your screen should look something like Figure 3-18.

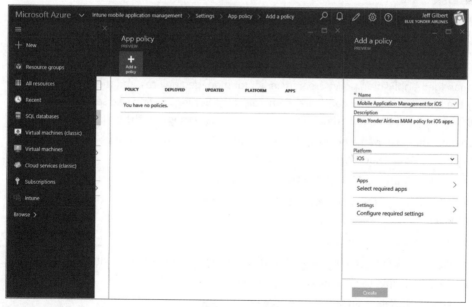

FIGURE 3-18 Creating a new MAM policy in the Azure portal

The policy now has a name and description, but you can't click Create until you select the apps the policy should apply to and configure the settings for the policy itself. To start this process, select Apps and then choose which apps to apply the new MAM policy to, as shown in Figure 3-19. In this case, you want to select all the currently available options, so press Control+Click to multiselect OneDrive, Excel, PowerPoint, and Word. Then click Select to complete the process.

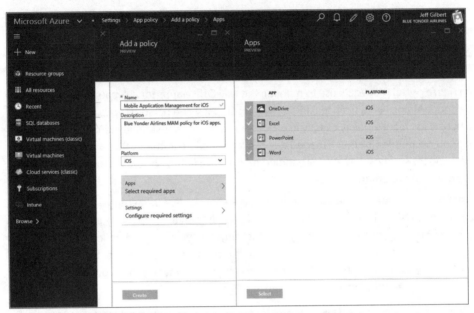

FIGURE 3-19 Selecting applications for a new MAM policy in the Azure portal

With the apps now selected for the policy, it's time to configure the actual policy settings themselves. The available policy-settings options are virtually identical to those you used for the managed-application policy created earlier for managed devices, and they perform basically the same functions. To configure them, select the Settings option to display the Settings blade, where you'll configure data relocation and access settings for the policy.

To meet the goals of the Blue Yonder Airlines MAM implementation strategy, you need to configure the Data Relocation settings as follows:

- Prevent iTunes and iCloud backups of app data.
- Allow managed apps to receive data from unmanaged apps, but not the other way around.
- Prevent "Save As" functionality.
- Restrict cut, copy, and paste to only other managed apps (with paste in).
- Force web links to open in a managed browser.
- Encrypt app data when the user's device is locked.

In other words, you want the Data Relocation settings section to look like the one shown in Figure 3-20.

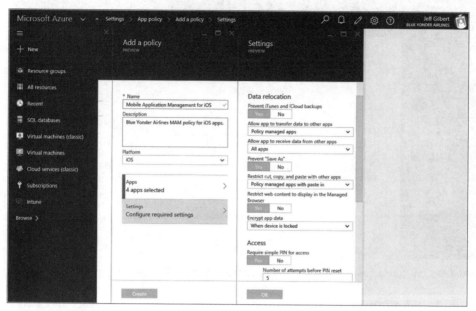

FIGURE 3-20 Configuring Data Relocation settings for a new MAM policy in the Azure portal

Moving down the blade from the Data Relocation section, you now need to finish configuring the Access settings for the policy. Again, this section is similar to the previous managed-application policy created in the Intune console, but there are a few differences.

Again, you'll require users to provide a simple PIN to access the managed applications, with five attempts allowed before a PIN reset is required. However, in this policy, you can also allow users to bypass the PIN and to use a fingerprint reader (iOS 8+) instead. You won't require them to use their corporate credentials to access the app, you'll continue to block jail-broken or rooted devices from running apps, and you'll maintain the same access requirements you set for the previous policy. Finally, you need to configure an offline interval, which specifies how many days that data for a company-managed app will remain on a user's device without being accessed before it is wiped. Accept the default value of 90 days, as shown in Figure 3-21.

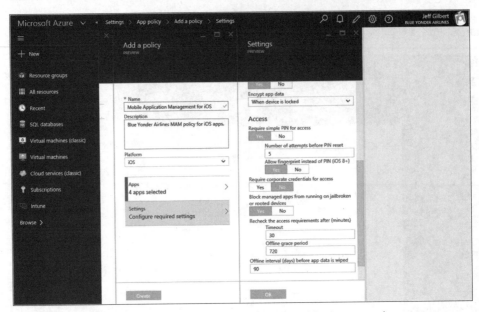

FIGURE 3-21 Configuring Access settings for a new MAM policy in the Azure portal

With all settings properly set, click OK to close the Settings blade and then click the Create button on the Add A Policy blade to finalize the new MAM policy. With your first Intune MAM policy created in the Azure portal, the policy is ready to be targeted to users and provide the data-leakage protection you desire for Blue Yonder Airlines.

At this point, you have completed most technical requirements for deploying managed iOS apps to Intune-managed devices and also prepared the Intune service using the Azure portal to manage applications on unmanaged devices. However, being the seasoned enterprise administrator that you are, you know there's still much work to be done before you can roll out these apps and polices to your users.

Performing the app and policy rollout

As the enterprise administrator for Blue Yonder Airlines, you're responsible for rolling out the needed Microsoft Office apps and related data-leakage protection policies. Those policies will ensure company employees can be productive on their devices regardless of whether or not they are managed and still keep company data secure. To accomplish this as quickly as possible, while also ensuring sustained adoption, you need to focus on organizational readiness and change management. You also need to consider the end-user experience of your rollout throughout the following phases:

- Project Scope
- Proof Of Concept
- Pilot

- Enterprise Rollout
- Run State

Scope the rollout project

So far, you have completed various planning and technical work, but it is during the Project Scope phase that you must ensure all pieces are in place to actually begin rolling out apps and policies to Blue Yonder Airlines employees. In addition to considering technical issues, you must now also consider the organizational issues involved.

Obviously, you cannot do everything on your own, so you must be sure to include other groups in your planning and scoping process. For example, you know you need security groups in Active Directory created and replicated into Azure Active Directory to target apps and policies to during the different phases of the rollout. You also know that you need to create groups of users and devices within the Intune administrator console to target app and policy deployments to.

During this phase, you need to crisply define what success looks like and agree on a timeline for the project. You need to be sure you understand exactly how the coming changes will affect the way users work and what needs to be done to support the long-term success of the rollout. Consider what training will be required for the help desk and end users and how you'll make that available. Finally, be sure that everyone from the CEO to the mail-room employees are aware of the overall scope of the project and kept informed throughout each phase of the rollout to avoid surprises.

After meeting with the Blue Yonder Airlines CEO and his leadership team, you agreed to an aggressive timeline to quickly complete this high-priority project. You also agreed to provide regular project status updates on measures of project success in meeting the following goals:

- Perform the app and policy rollout without disrupting day-to-day work of Blue Yonder Airlines employees.
- Microsoft Office mobile applications and the Intune Managed Browser are successfully deployed on iOS devices.
- If an Office app is already installed, you must successfully take over management of the app.
- Implement mobile application management policies to manage app and data behavior on both managed and unmanaged devices.
- Regularly report on the app-installation status.

For this project, the Chief Technology Officer (CTO) is the executive sponsor, and he and his team will manage the awareness and communications plan for Blue Yonder Airlines employees based on your project-status updates. Adam Barr, one of the other senior admins on the team, will work with the help desk to ensure they are ready to support the coming changes and also develop some in-house training sessions for users, which will be posted on the company intranet.

Proof of concept

With the project properly scoped, you can move on to the Proof of Concept phase, where you'll perform lab testing to understand and verify the viability of the planned deployment scenarios in your environment. During this phase, you'll deploy the previously created apps and data-leakage protection policies to test devices to validate your plan and ensure you can meet the goals of the overall rollout project.

Deploy apps and policies without disrupting employees

Remember that one of the overarching goals of this project is to increase, not decrease, the productivity of Blue Yonder Airlines employees. Even though end users will not be asked to do much during the rollout, you should ensure they are expecting the changes you'll be pushing to their devices. They should expect, and trust, that the apps they want and need are going to be installed in a thoughtful manner, and they should know how these apps will work and will benefit them.

Although much of the work for this goal will land with Adam as part of his end-user training and awareness campaign, you should consider the technical aspects involved. You need to ensure users are made aware of any tasks they need to perform during app installations and informed of a timeline for the changes to happen on their device. You know that iOS devices check in with the Intune service every six hours for policy changes, so you let Adam know to plan to give users a day-long window for the installations to occur once production deployments begin.

Successfully deploy Microsoft Office mobile apps for iOS and the Intune Managed Browser

To meet this goal during the Proof of Concept (POC) phase, you need to deploy your previously created managed iOS apps to actual iOS devices. For this small assessment lab deployment, you select a few iOS test devices (a few iPads and one iPhone) and a small, POC test-user group to deploy the applications to.

To kick off this phase, you log in to the Intune administrator console and navigate to the APPS workspace. Navigate to the Apps page, and finally double-click the Microsoft PowerPoint app to view the first app to be test deployed, as shown in Figure 3-22.

FIGURE 3-22 The Microsoft PowerPoint app for iOS

Click Manage Deployment to configure the deployment settings for the Microsoft Pow-erPoint app for iOS. Next, select the Proof Of Concept Group from the list of user groups you created during the project's scoping phase, as shown in Figure 3-23.

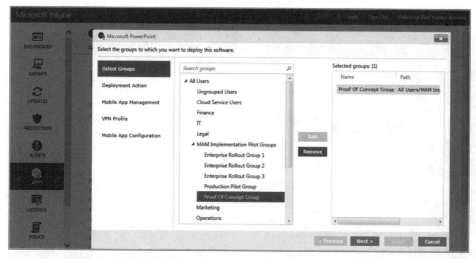

FIGURE 3-23 Selecting the Proof Of Concept Group to target deployment of the Microsoft PowerPoint app for iOS

After selecting the appropriate group to target the app deployment to, you configure the Approval setting to require the installation and the Deadline setting for that installation to occur as soon as possible, as shown in Figure 3-24.

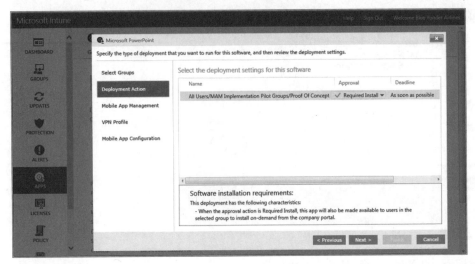

FIGURE 3-24 Configuring the Approval and Deadline options for the Microsoft PowerPoint app for iOS

Moving on to the Mobile App Management settings, you now need to associate the mobile application management policy you created for managing applications deployed to managed devices with this deployment. To do that, you select the policy name from the App Management Policy drop-down list, as shown in Figure 3-25.

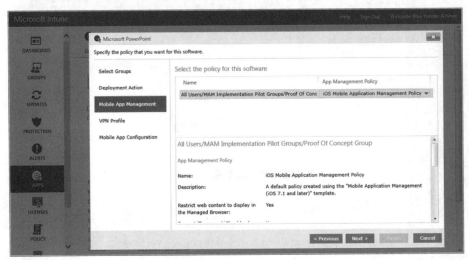

FIGURE 3-25 Configuring the Mobile App Management settings for the Microsoft PowerPoint app for iOS

Because you do not need a virtual private network (VPN) profile to access app data and you will not be using a mobile app configuration for this app, you click Next and then Finish to close the app properties and complete the deployment of the Microsoft PowerPoint app for iOS to the Proof of Concept user group. The app will now reflect a deployed status in the Intune administrator console, and app installations will soon begin.

Go to your iPad test device, log in to the Intune Company Portal App with lab test credentials for a user in the Proof of Concept user group, and force a quick policy synchronization. Within seconds, you're prompted to install the managed PowerPoint app from the App Store. Click Install, and the app begins loading and installs successfully. Continuing, repeat this process for the remaining Microsoft Office for mobile apps and verify that they all install correctly with no issues.

The final app to deploy is the Microsoft Intune Managed Browser app. Because this app is not a managed iOS app, you will not be able to make the deployment mandatory for the test users. Instead, you set the approval setting for the deployment to Available Install and make a note to be sure Adam informs users they need to install the Intune Managed Browser and explains how to do so. Then you click Finish, as shown in Figure 3-26.

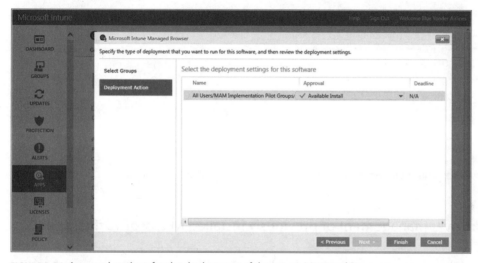

FIGURE 3-26 Approval settings for the deployment of the Intune Managed Browser app

After you click Finish, the app deployment request is quickly processed and the app is successfully deployed to the Proof of Concept user group. Back on your test device, you now see the Managed Browser app available to install from the App Store. Click the link to install the Microsoft Intune Managed Browser from the Apple App Store, and verify that it opens successfully, as shown in Figure 3-27.

FIGURE 3-27 The Intune Managed Browser app successfully installed on an iOS device

Take over management of previously installed Microsoft Office apps on iOS

You confirmed the expected behavior for apps deployed to devices that did not already have them installed. Now you need to double-check that the user experience works as expected for taking management ownership of an iOS app that already has been installed by the device user.

The easiest way to test this behavior with a lab device is to complete the following steps:

1. Manually install the app directly from the Apple App Store on a freshly wiped iOS device.

2. Install the Intune Company Portal app.

3. Enroll the device into management with Intune using Proof Of Concept user account credentials.

After you enroll the device into management, you should see a message asking you to allow Intune to take over management of the previously installed Microsoft PowerPoint app, as described in Chapter 2. Select Manage in the app management change dialog box to confirm the process is working as expected and allow Intune to manage the app.

Successfully implement mobile application management policies

With the app installation behaviors validated, the next testing scenario to validate is whether or not the mobile application management policies are working correctly. For this scenario, you need to test for MAM policy effectiveness on both managed and unmanaged iOS devices.

Starting with a managed iOS device, open the Microsoft PowerPoint managed app. You should be prompted to set a numerical PIN based on the MAM policy settings you previously configured, as shown in Figure 3-28.

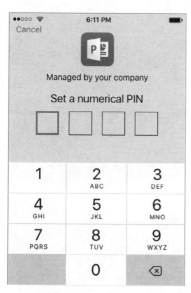

FIGURE 3-28 The managed Microsoft PowerPoint App for iOS prompting for a PIN

With the access PIN requirement working as expected, you next need to check the data-leakage policy options to restrict cut, copy, and paste to only managed apps. This time, you open Microsoft Word, provide the access PIN (which is the same as the one you created earlier for PowerPoint because all Microsoft apps using this policy will use the same PIN), and write a few simple sentences to check for policy enforcement, similar to what is shown in Figure 3-29.

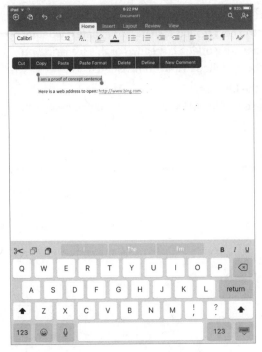

FIGURE 3-29 Testing data-leakage protection policies with Microsoft Word for iOS

To validate the policies are working as expected, you attempt to copy and paste the test sentence into an unmanaged app, such as notes, on the test iPad. When you do so, the option to paste is not enabled, but when you attempt to paste the test sentence into the managed Microsoft PowerPoint app, the pasting function works. Back in the test Word document, clicking the web address opens *http://www.bing.com* in the Intune Managed Browser as expected, as shown in Figure 3-30.

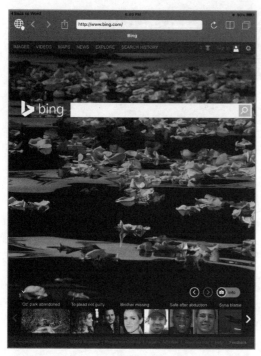

FIGURE 3-30 The Microsoft Intune Managed Browser app opening bing.com

With the MAM policy settings verified for the apps deployed by Intune to managed devices, the only thing left to do now is confirm that the policy is working as expected on unmanaged devices.

Just like you had to deploy apps and associated policies to users in the Intune console, you also need to deploy the MAM policy you created in the Azure portal for apps on devices not managed by Microsoft Intune. To do that, you log in to the Microsoft Azure portal using your Microsoft Intune admin credentials, and deploy the previously created MAM policy to the group of users that you want the policy to apply to. In this case, you navigate to the MAM policy's User Groups settings and select the Azure-based Managed Microsoft Office Users group you created earlier, as shown in Figure 3-31.

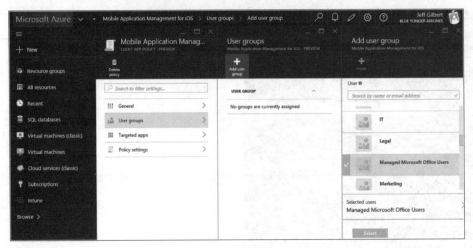

FIGURE 3-31 How to specify user groups to apply MAM policies to through the Azure portal

After saving the policy, use a recently wiped and reset iOS test device to validate that the MAM policy settings configured in the policy are working as expected. Remember that when using MAM policies for unmanaged devices, you (and your end users) must install all the apps to be managed manually from the Apple App Store.

Continuing your testing, and using a test lab user credential that is not part of the Proof of Concept user group configured in the Microsoft Intune console, you log in to an unmanaged iOS device and install the apps to be managed as well as the Microsoft Intune Managed Browser. Using your test Blue Yonder Airlines credentials to log in to the Microsoft Office apps results in those apps becoming managed by the MAM policies you configured in the Microsoft Azure portal. A message similar to that shown in Figure 3-32 alerts users that their applications are now being centrally managed.

FIGURE 3-32 MAM policies configured through the Azure portal taking effect

Because the MAM policy settings you configured in the Azure portal are basically the same as those you configured from within the Intune administrator console, the app and data behavior in your proof-of-concept testing should proceed along the same lines as the earlier testing.

Report on app installation status

Microsoft Intune also provides you with several reporting options you can use to regularly check on the progress of app installations. During the Proof of Concept stage, you'll verify that the reporting information available will suffice to keep everyone informed throughout the lifetime of the app rollout project.

The easiest way to check on any Microsoft Intune app deployment is to filter the view on the Apps page of the APPS workspace in the Microsoft Intune administrator console. From there, you can review the deployed app-user or device summary to quickly view the number of app installations or failures for all deployed apps in a single glance.

If you're looking for more information about a single app deployment, you need only highlight the app title to quickly gain an understanding of how many users or devices have the app installed or available, as well as whether there are any deployment failures to be investigated. You can see both views displayed in Figure 3-33.

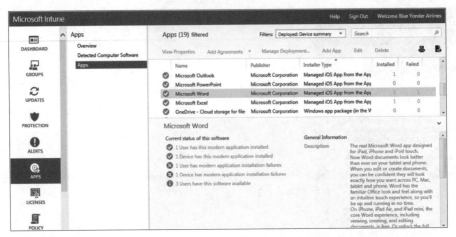

FIGURE 3-33 Monitoring app deployment status in the Intune administrator console

With the technical deployment and policy settings validated successfully in a small test lab environment, and all project stakeholders informed of your results, you're ready to begin the next phase of the project, the production Pilot phase.

Small production Pilot phase

While all testing during the proof of concept phase occurred on test devices and without any real risk of data loss, the production pilot will involve real company data and devices used by actual Blue Yonder Airlines employees. During this phase, you continue to validate the technical capabilities and features you previously implemented, but you also test more complicated scenarios in a limited production environment.

The production Pilot phase for Blue Yonder Airlines consists of a small group of early-adopter and technology-savvy end users who the team identified during the Project Scope phase. For the production Pilot, those users will have managed apps and policies deployed to their devices and they'll provide feedback on the usability and effectiveness of the policies. In addition to the technical aspects of the Pilot phase, these users will also be the first to receive rollout awareness communications and evaluate the end-user training materials for effectiveness. The help desk will also be involved to provide proactive assistance to ensure all Pilot participants are fully supported throughout the Pilot testing period.

At the end of the production Pilot, any technical issues that users encountered can be remediated and the user experience can be improved before the larger, enterprise rollout even begins.

Enterprise Rollout phase

The Enterprise Rollout phase is the largest deployment phase and encompasses all aspects of the project to date. It builds on the successes and lessons from the earlier deployment phases to ultimately deploy apps and policies to a much broader set of users and groups. These users will all require an awareness of upcoming changes through clear communications, training on what to expect, and expert support from the help desk.

For the Blue Yonder Airlines Enterprise Rollout phase, you'll target specific groups of users based on their location to perform a phased rollout of the apps and policies over a period of weeks. By using this approach, you can make adjustments to the plan or schedule if the need arises and ensure a positive first experience for all employees.

During this phase, it is critical to remain vigilant and monitor app deployments, quickly identify and troubleshoot any app installation failures, and continue to communicate freely with your rollout partners. Continue to report project status and progress against goals to all members of the project team until you complete the rollout.

Run State phase

After the enterprise rollout is completed, all Blue Yonder Airlines employees using iOS mobile devices have a managed version of Microsoft Office installed and company data is protected whether or not the device itself is managed by Microsoft Intune.

Even though the initial rollout is completed, you must continue to work to ensure the long-term success and sustained adoption of the Microsoft Intune mobile application management capabilities you have enabled. Even though all current employees are now trained and using mobile apps to stay productive, you must ensure a process is in place to retain the rollout momentum so that future hires for the company can be onboarded easily and the help desk stays up to date with the latest application and policy requirements.

Congratulations! With the Run State phase completed and deployment goals met, you have successfully implemented your Microsoft Intune mobile application management strategy for Blue Yonder Airlines.

CHAPTER 4

Introducing Microsoft Advanced Threat Analytics

When enabling enterprise mobility in a hybrid environment where cloud services are integrated with an on-premises infrastructure, you need to ensure that on-premises security controls are in place to protect company data. The problem with security controls that are currently implemented in some environments is the limited protection provided when sophisticated security breaches occur or when a user's identity is compromised. Although Microsoft Azure Active Directory Premium reports can help to identify suspicious activities around authentication, organizations need to have a mechanism to monitor on-premises resources and mitigate potential attacks.

Microsoft Advanced Threat Analytics (ATA) can help companies identify suspicious user and device activity. By leveraging advanced machine-learning technology, ATA helps identify suspicious user and device activity while providing clear and relevant attack information. This chapter explains this new member of the Enterprise Mobility Suite (EMS) and covers the design considerations that should be taking place before implementing it.

Protecting on-premises resources

Although there is a trend to move resources to the cloud, most environments are still in transition, which means they are in a hybrid mode. While in hybrid mode and adopting mobility, you must be fully aware of your threat landscape. However, being aware is not enough; actions must be taken to mitigate attacks in a world of constantly evolving cyber tactics. IT has to adapt as fast as the attackers. IT needs to identify abnormal behavior and take actions to either mitigate a potential threat or ignore it in cases of expected behavior for their environment. On-premises security becomes particularly important in "bring your own device" (BYOD) scenarios, where users are bringing their own devices to an on-premises infrastructure.

To enhance the security of your on-premises infrastructure, Microsoft includes ATA as part of the EMS solution. The goal is to have an end-to-end solution that carries security considerations for data and user identity throughout the entire communication channel, from the device, to the cloud, and to the on-premises datacenter, as shown in Figure 4-1.

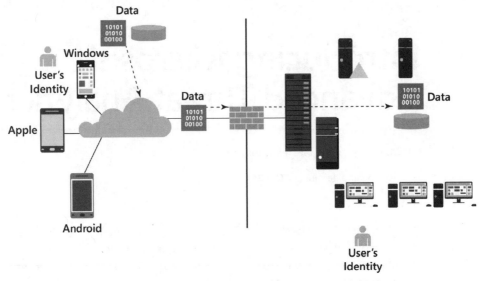

FIGURE 4-1 End-to-end protection of data and user identity is a requirement for enterprise mobility

Understanding ATA

Microsoft ATA uses machine learning for analyzing entity behavior. Using deep packet inspection technology, ATA analyzes all on-premises Active Directory network traffic. After analysis, ATA builds an Organizational Security Graph, a living, continuously updated view of all users, devices, and resources within an organization that understands normal network behavior. ATA looks for any abnormalities in the entities' behavior and raises alerts, but not before abnormal activities have been contextually aggregated and verified.

> **MORE INFO** For more information about entity relationship behavior, read *https://msdn. microsoft.com/en-us/library/gg309412.aspx*. For more information about deep packet inspection, read *http://www.techrepublic.com/blog/data-center/deep-packet-inspection- what-you-need-to-know*.

ATA analyzes the behaviors among users, devices, and resources, and their relationship to one another. It can also detect known attacks. ATA contextually aggregates suspicious activities before alerts are issued. What ATA detects is based on three variants:

- Abnormal user behavior, such as *anomalous login*.
- Malicious attack, such as *pass-the-hash*. For more information about pass-the-hash attacks, read *https://technet.microsoft.com/en-us/dn785092.aspx*.
- Known security issues and risks, such as exploitation of weak protocols.

IMPORTANT ATA can also collect relevant events from Security Information and Event Management (SIEM) systems or from domain controllers via Windows Event Forwarding (WEF). In summary, the sources of information for ATA are domain controller network traffic, Active Directory itself (via LDAP queries), and Windows event logs (including security event 4776, SIEM, and WEF).

Real World   Why deploy ATA on-premises?

It seems that almost every week another company or organization ends up in the news, and not because they released a new product or had amazing financial results. Rather, they have been hacked, and sensitive and confidential information has been released or the perpetrators have threatened to release it. These companies are in all industries: retail, entertainment, financial, online services, medical, cyber security, and others. No sector has been spared. It even happens to governments.

Many hacked companies spent lots of money on the best security technology available, yet they were stilled hacked. Part of the reason is that the security products deployed were designed to protect the internal network from external attacks from the Internet; they provide *perimeter* protection. These perimeter-protection technologies were not designed to protect the internal on-premises network and resources from internal attacks. Additionally, because hackers spend more and more time in an organization's network and use everyday tools that are not detected by current tools, the protection paradigm has shifted to assumed breach[1]. Now organizations need to be also looking on the internal network for abnormal user and entity behavior.

As organizations move to more cloud services, user identities and on-premises resources still need to be monitored and protected.

ATA analyzes and learns user and entity behavior by aggregating data from multiple data sources, such as deep packet inspection of domain-controller traffic, Windows events, and data provided via integration with third-party solutions. Using its proprietary algorithm, ATA works around the clock to help you pinpoint suspicious activities in your systems by profiling and knowing what to look for. ATA also identifies known advanced attacks and security issues.

The ATA attack timeline offers a clear, efficient, and convenient social network–like feed for suspicious activities. The timeline gives you a powerful perspective on the "who, what, when, and how" in your enterprise. ATA also provides recommendations for investigation and remediation for each suspicious activity.

Gershon Levitz
Senior Program Manager, Microsoft Advance Threat Analytics Team

[1] Read more about assume breach methodology in this paper: http://download.microsoft.com/download/C/1/9/ C1990DBA-502F-4C2A-848D-392B93D9B9C3/Microsoft_Enterprise_Cloud_Red_Teaming.pdf.

To detect common attacks, ATA parses all the network traffic that comes from domain controllers, and then resolves the entities involved in a logical representation of the traffic. Once this process is done, results are sent to the ATA Center, where the parsed traffic will be stored and all detection engines will be run on the database. ATA adapts to changes, identifies abnormal behavior with its proprietary algorithm, and reports anomalies. Once ATA is deployed on-premises, it continually monitors the on-premises infrastructure to ensure protection based on the following four phases:

- **Analyze** In this phase, ATA uses preconfigured, nonintrusive port mirroring. All Active Directory–related traffic is copied to ATA while remaining invisible to attackers. ATA uses deep packet inspection technology to analyze all Active Directory traffic.

- **Learn** ATA starts learning and profiling behaviors of users, devices, and resources, and then leverages its self-learning technology to build an *Organizational Security Graph*. This graph is a map of entity interactions that represent the context and activities among users, devices, and resources.

- **Detect** In this phase, ATA looks for any abnormalities in an entity's behavior and identifies suspicious activities.

- **Alert** If abnormal or suspicious activities occur, ATA triggers an alert. For each suspicious activity, ATA provides recommendations for the investigation and remediation at the ATA Console.

To further increase accuracy and save time for IT administrators, ATA doesn't only compare the entity's behavior to its own, but also to the behavior of other entities in its interaction path before issuing an alert. This means that the number of false positives are greatly reduced. To better understand how ATA works, you need to know which components are included in a common on-premises topology. Figure 4-2 shows a standard topology with ATA components.

There are two core components for ATA deployment: the ATA Gateway and the ATA Center. The *ATA Gateway* is responsible for analyzing network traffic to and from domain controllers, receiving data from entities located in the on-premises domain, receiving information from the SIEM system (if present), and transferring data to the *ATA Center*. All administration occurs via a web-management interface on the ATA Center, which is responsible for managing the ATA Gateway settings and detecting suspicious activities and abnormal behavior.

Communication between the ATA Center and the ATA Gateway is encrypted using Secure Sockets Layer (SSL) on TCP port 443. For this configuration to work properly, you need to configure two IP addresses on the ATA Center. The ATA Center service will bind port 443 to the first IP address and IIS will bind port 443 to the second IP address. Although you can use a single IP address and configure two different ports, having two IP addresses is the recommended setup.

ATA requires that you configure port mirroring in the network switch where the ATA Gateway is connected with the domain controllers. This requirement is necessary to allow ATA to perform deep packet inspection on traffic to and from the domain controllers to identify known attacks. ATA also uses network traffic to learn which users are accessing which resources and from which

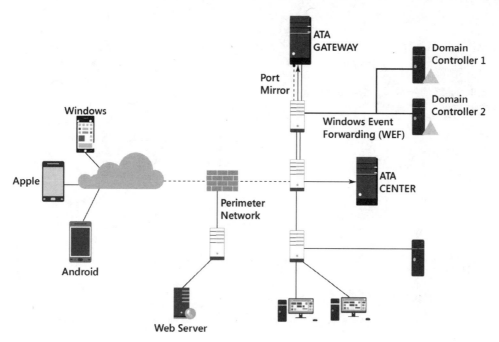

FIGURE 4-2 ATA components distributed in an on-premises infrastructure

computers. In addition, ATA makes LDAP queries to domain controllers to create user and device profiles, using a custom user account that requires read-only access to the domain.

> **IMPORTANT** One ATA Center can support multiple ATA Gateways. For more information about sizing, read the "Planning and designing ATA" section later in this chapter.

With ATA, there isn't a need to create rules, thresholds, or baselines and then fine-tune each at a later time. Approximately three weeks after deployment, ATA starts to detect behaviorally suspicious activities. However, ATA will start detecting known malicious attacks and security issues immediately after deployment.

Deploy ATA in a restrictive communication environment

In some circumstances, you might need to deploy ATA in an environment that has a firewall that controls the communication between subnets or even within the same subnet. When the ATA Center and Gateway are in between a network device that filters the communication, such as a firewall, you will need to open ports to allow this communication to take place. The scenario shown in Figure 4-3 has the ATA Center isolated in a protected network. Use Table 4-1 to understand which ports should be open, the direction, and the purpose.

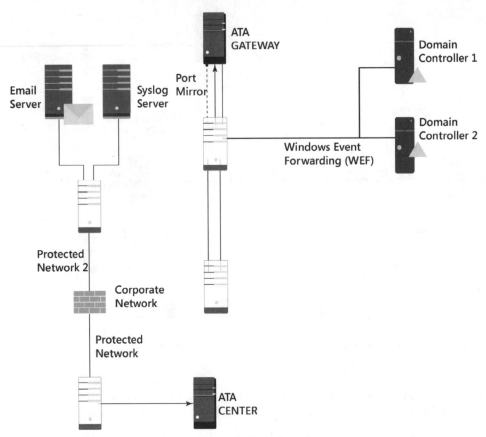

FIGURE 4-3 ATA Center located in a security zone behind a firewall

TABLE 4-1 Protocols and ports for ATA Center communications

Application	Transport protocol	Port	To/From	Direction	IP Address
SSL (ATA Communications)	TCP	443 (Default)	ATA Gateway	Inbound	IP 1
HTTP	TCP	80	Company Network	Inbound	IP 2
HTTPS	TCP	443	Company Network and ATA Gateway	Inbound	IP 2
SMTP (optional)	TCP	25	SMTP Server	Outbound	IP 2
SMTPS (optional)	TCP	465	SMTP Server	Outbound	IP 2
Syslog (optional)	TCP	514	Syslog server	Outbound	IP 2

Another common scenario is when the ATA Gateway is located in a network behind a firewall, as shown in Figure 4-4. In this scenario, use Table 4-2 to understand which ports should be open, the direction, and the purpose.

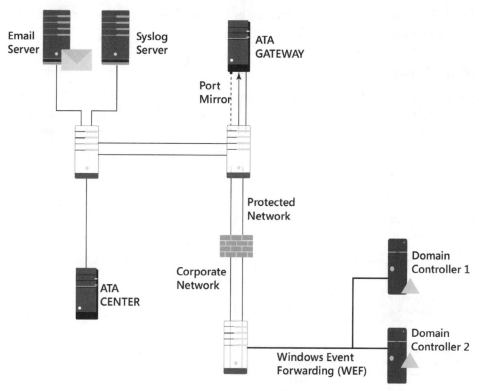

FIGURE 4-4 ATA Gateway located in a security zone behind a firewall

TABLE 4-2 Protocols and ports for ATA Gateway communications

Application	Transport protocol	Port	To/From	Direction
SSL	TCP	443	ATA Center (both IPs)	Outbound
LDAP	TCP and UDP	389	Domain controllers	Outbound
LDAPS	TCP	636	Domain controllers	Outbound
LDAP to Global Catalog	TCP	3268	Domain controllers	Outbound
LDAPS to Global Catalog	TCP	3269	Domain controllers	Outbound
Kerberos	TCP and UDP	88	Domain controllers	Outbound
Netlogon	TCP and UDP	445	Domain controllers	Outbound
Windows Time	UDP	123	Domain controllers	Outbound
DNS	TCP and UDP	53	DNS Servers	Outbound
NTLM over RPC*	TCP	135	All devices on the network	Outbound
NetBIOS*	UDP	137	All devices on the network	Outbound

As part of the name-resolution process done by the ATA Gateway, these ports need to be open inbound on devices on the network from the ATA Gateways.

ATA architecture

As mentioned previously, ATA is distributed in two key roles (Gateway and Center). These roles have different tasks to perform, and their core architectures are slightly different. The most complex architecture is the ATA Gateway. Figure 4-5 shows the user-mode elements that are used by ATA Gateway.

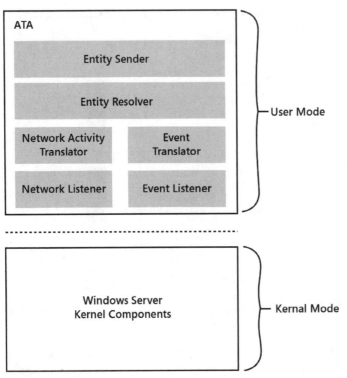

FIGURE 4-5 Core Microsoft ATA architecture

You can see that all core elements of ATA run in user mode and each has a unique task. The following list provides a brief explanation of each element:

- *Entity Sender* is responsible for sending the parsed and matched data to the ATA Center.
- *Entity Resolver* takes the parsed data (network traffic and events) and resolves the data with Active Directory to find account and identity information and match it with the IP addresses found in the parsed data. In addition, the Entity Resolver will also inspect the packet headers, enable parsing of authentication packets for machine names and properties, and identify and combine all these attributes with the data in the actual packet.

- *Network Listener* is responsible for capturing network traffic and parsing the traffic. This is a CPU-intensive task, so you need to plan and design the solution well.
- *Event Listener* is responsible for capturing and parsing Windows events forwarded from a syslog server (in case one exists in your infrastructure).
- *Network Activity Translator and Event Translator* are responsible for converting raw network traffic and events into an ATA-logical representation of them.

Enhance enterprise mobility security with ATA

According to Verizon Data Breach Investigation Report 2013, 76 percent of all network intrusions occur because of compromised user credentials. When users carry their mobile devices and use cloud apps, the threat landscape expands and the likelihood of data being compromised increases. If your company doesn't have a system that monitors users' behavior, it will be hard to identify users who are bringing a compromised device on-premises to access resources in the corporate network. Cyber security attacks are evolving, and in some cases it can take up to eight months before a breach is detected.

> **MORE INFO** For more information, read *http://www.verizonenterprise.com/resources/reports/rp_data-breach-investigations-report-2013_en_xg.pdf*. Read this article for more information: *http://blog.trendmicro.com/catch-me-if-you-can-how-apt-actors-are-moving-through-your-environment-unnoticed/*.

ATA closes this gap for the enterprise mobility solution by leveraging all capabilities that were previously explained to identify potential threats. For companies embracing enterprise mobility, this is a critical step toward a safer deployment as employees use different devices and are exposed to more diverse threats. The fictitious company Blue Yonder Airlines (described in Chapter 1, "Understanding Microsoft enterprise mobility solutions") will leverage ATA to monitor on-premises resources and identify abnormal behavior in the network.

Planning and designing ATA

Because ATA is a product that is installed on-premises, you must be aware of the prerequisites from the infrastructure perspective and from the server-side perspective. You must understand how to properly design the solution prior to deployment to start using it effectively.

Infrastructure considerations

You should review your current on-premises infrastructure prior to deploying ATA. Ideally, you should already have a good understanding of what expected network behavior is, knowing network peak times and the core protocols that are used in transit. From the networking perspective, you need to identify the subnets where the DHCP-managed IP address leasing has a short Time To Live (TTL) duration (seconds or minutes). This is an important consideration because, during the ATA installation, you can make the appropriate changes to the short-term IP address lease configuration to force ATA to reduce the cache lifetime for all IP addresses in that particular subnet (or subnets). This change is key to better accommodating the fast reassignment between devices. Evaluate not only the local area network (LAN) infrastructure, but also virtual private network (VPN) and Wi-Fi networks because usually these subnets have short-term IP address lease durations. Also, make sure to evaluate your network infrastructure to understand where the ATA Gateway and ATA Center will reside.

Before deploying ATA, ensure that your on-premises infrastructure has domain controllers running Windows Server 2008 or later, and that there is a user account created with read access in the domain that will be monitored by ATA. If your organization has an Active Directory infrastructure that has multiple organizational units (OUs) with a different access control list for each OU, make sure this account will have at least read access to all these OUs. Also keep in mind that you can deploy only one ATA instance per forest, even if this forest has multiple domains. If your environment consists of multiple forests, you will need multiple ATA deployments.

Port mirroring must be configured for each domain controller. This step is mandatory so that ATA can see all traffic to and from each domain controller. Each domain controller is configured as the source of the port mirroring, and the ATA is the target. This configuration can be done in a physical or virtual switch.

> **IMPORTANT** From the supportability standpoint, some combinations of virtual and physical port mirroring aren't supported. For more information, read the supported port mirror options in "Configure Port Mirroring" at *https://technet.microsoft.com/en-us/library/mt429376.aspx*. Also note that you can't have a domain controller running in Azure Information as a Service (IaaS) because port mirroring isn't supported there.

You must also take into consideration the number of domain controllers you have and the traffic that will be generated. One ATA Gateway can support multiple domain controllers; however, you must understand your current environment to correctly size the amount of ATA Gateways needed for each domain controller. Use the guidelines from *http://aka.ms/atasizing* to correctly size your ATA solution prior to deployment.

ATA Center considerations

To install the ATA Center, you must first have a server with Windows Server 2012 R2 (workgroup or in a domain-joined environment) configured with all current operating-system updates. The installation can be done in a physical server or in a virtual machine. The minimum hardware requirements are:

- CPU: Eight cores
- Memory: 48 GB

- Storage: 1000 GB per month to monitor two light domain controllers
- Network: One network adapter with two IPs

You can either install a self-signed certificate during the installation of the ATA Center or install a certificate from an internal or external public Certification Authority (CA) like VeriSign. If you decide to use a self-signed certificate during the installation, you can replace it for another one post installation. The certificate will be used by IIS and for server authentication purposes.

ATA Gateway considerations

From the operating system perspective, the ATA Gateway requirement is the same as the ATA Center. You can install ATA Gateway in workgroup or in a domain-joined environment. However, you can't install it on a domain controller. The installation can be done on a physical server or a virtual machine. The minimum hardware requirements are as follows:

- **CPU:** Four cores
- **Memory:** Eight GB
- **Storage:** Enough for the OS + 10 GB for ATA + crash dumps = at least 100 GB
- **Network:** Two network adapters; one for management and one to capture traffic from the domain controller

> **IMPORTANT** The time synchronization between the ATA Center server and the ATA Gateway server must be within five minutes of each other.

Just as we recommended for the ATA Center, here you can also use a self-signed certificate during the installation of the ATA Center or install a certificate from an internal or external public Certification Authority (CA) like VeriSign. If you decide to use a self-signed certificate during the installation, you can replace it for another one post installation.

ATA Console considerations

The access to the ATA Console is done via a web browser. (Internet Explorer 10 or Chrome 40 are the minimum requirements.) The minimum screen resolution required is a 1700-pixel resolution.

CHAPTER 5

Implementing Microsoft Advanced Threat Analytics

I n Chapter 4, "Introducing Microsoft Advanced Threat Analytics," you learned how Microsoft Threat Analytics (ATA) can be used to help meet the challenges of emerging threats. You learned about the requirements for the Gateway and Service server role in the ATA architecture and general planning considerations. To keep company data secure in the cloud and on the premises, you must mitigate potential threats before they compromise your systems. In a hybrid scenario, interaction with on-premises resources is common, and companies embracing enterprise mobility need to make it a priority to identify suspicious activities and properly respond to a security incident.

In this chapter, you adopt the persona of the senior enterprise administrator for Blue Yonder Airlines and work to address the on-premises protection requirements described at the end of Chapter 1, "Understanding Microsoft enterprise mobility solutions." Remember that in this phase you're responsible for preparing and successfully implementing Microsoft ATA for the company.

Scenario requirements for on-premises protection

As a senior enterprise administrator for Blue Yonder Airlines, you are responsible for planning, designing, and implementing the company's enterprise mobility management (EMM) solution. In this phase of the implementation, Blue Yonder Airlines needs to enhance the overall security of its on-premises resources by installing and configuring Microsoft ATA.

Implementation goals

Blue Yonder Airlines has already reviewed all requirements to implement ATA, and Microsoft Windows Server 2012 R2 servers are installed and running the latest updates. The entire infrastructure was reviewed, the Internet Protocol (IP) used by each server is already reserved, and port mirroring is configured in the switch where the domain controller is connected. A read-only user is created in Active Directory to be used by the ATA Gateways.

As a senior enterprise administrator for Blue Yonder Airlines, you're ready to start the implementation. At the end of this phase, the following goals must be accomplished:

- Monitor on-premises resources, and identify abnormal behavior in the network.
- Detect attacks that exploit advanced tools and techniques in your on-premises environment.
- Detect security issues and risks, and alert administrators that they are happening,
- Reduce false-positive alerts to avoid creating unnecessary red flags and distraction from real issues.

Solution diagram

To meet the Enterprise Mobility Suite (EMS) implementation goals for the second phase of the EMS project, you'll implement the solution shown in Figure 5-1.

> **TIP** This solution diagram provides a high-level overview and basic description of the intended solution architecture.

You have some important considerations regarding this solution diagram:

- Because of some constraints regarding the servers that should be domain-joined, Blue Yonder Airlines decided to not join the ATA servers to the corporate domain.
- ATA servers are on the same subnet segment as the domain controllers.
- The workstation shown in the diagram is the one that will be used for administrators to connect via remote desktop to ATA Center/Gateway.

Deploying ATA

ATA deployment consists of installing the two server roles: the ATA Center and the ATA Gateway.

> **TIP** If you want to evaluate Microsoft ATA before deploying it in a production environment, download the 90-day trial at *https://www.microsoft.com/en-us/evalcenter/evaluate-micro-soft-advanced-threat-analytics*.

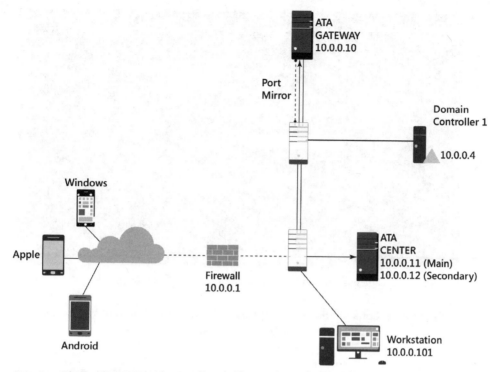

FIGURE 5-1 Blue Yonder Airlines solution architecture diagram

Installing ATA Center

To start the installation of ATA Center, ensure that you have the Microsoft ATA DVD media inserted or the ISO mounted in the server. After you finish the ATA Center installation, the following components will be installed and configured on the server:

- Internet Information Services (IIS)
- MongoDB
- ATA Center service and ATA Console IIS site
- Custom Performance Monitor data collection set
- Self-signed certificates (if you selected this option during the installation)

Sign in to this server with an account that has local administrative privileges, and complete the following steps to start the installation:

1. Launch the Microsoft ATA Center Setup. On the Welcome page shown in Figure 5-2, select your preferred language and click Next.

2. Read the Microsoft Software License Terms, select I Accept The Microsoft Software License Terms to agree, and click Next to continue.

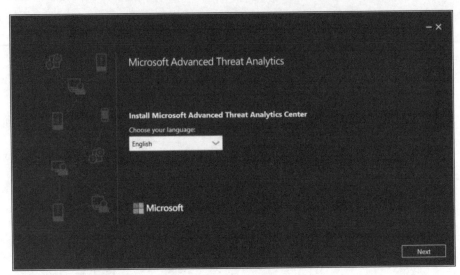

FIGURE 5-2 Language selection available during Microsoft ATA Center setup

3. The ATA Center Configuration page appears as shown in Figure 5-3. This is an important setup option. You can customize how your ATA Center is configured, which has a direct impact on the system's performance.

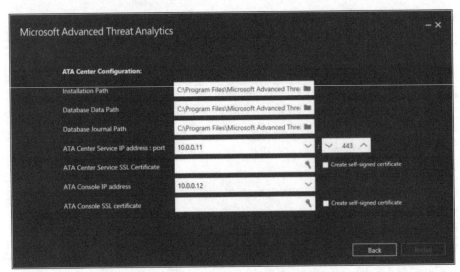

FIGURE 5-3 You use ATA Center configuration to customize the location of critical files

Consider the following recommendations and issues before making changes to these options:

- For large deployments (more than 100,000 packets per second), it's required that the database journal be located on a different disk then the database data.

- Ideally, you should have a dedicated disk for data. If you do not, ensure that the disk where the database data is going to be located has more than 20 percent of free space. Be aware that if the disk's free space reaches a minimum of either 20 percent or 100 GB, the oldest 24 hours of data will be deleted.

- You can create self-signed certificates to be used during the installation and later replace it with a certificate from an internal Certification Authority (CA) to be used by the ATA Gateway. The ATA Center Services Certificate is the certificate used by the ATA Center service and the ATA Console SSL certificate is used by IIS.

- Ensure that the ATA Center Services IP (the first IP bound to the interface) and port are correct before proceeding. This is the IP that listens for communication from the ATA Gateways.

- Ensure that the ATA Console IP (the secondary IP in the same network interface card) address is correct. This is the IP address that is used by IIS for the ATA Console.

4. For this installation, select Create Self-Signed Certificate for both of the following options: ATA Center Services SSL Certificate and ATA Console SSL Certificate. Leave the other options as they are, and click Install. The progress page, shown in Figure 5-4, appears and requests that you restart the system when the installation is finished.

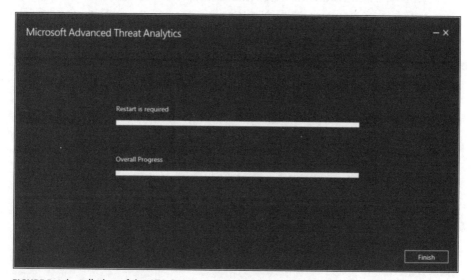

FIGURE 5-4 Installation of the ATA Center was successfully completed

5. Click Finish to proceed, and click Restart Now in the dialog box that appears.

6. After the server restarts, sign in with the same account with which you started the installation. You will notice that the setup resumes and continues the installation, as shown in Figure 5-5.

7. After the setup process has completed, the Launch button is available in the right corner of the window. Click it to launch Microsoft ATA Center.

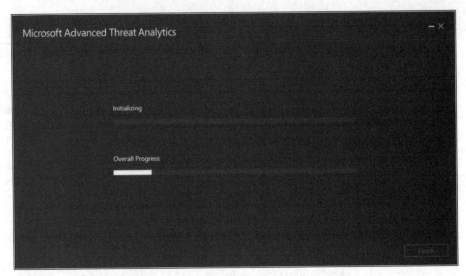

FIGURE 5-5 The ATA Center setup process resumes after the server restarts

8. When you launch the ATA Console, Internet Explorer opens. Because the IP address of the site does not match the certificate's subject name, Internet Explorer will exhibit a warning similar to the one shown in Figure 5-6.

FIGURE 5-6 The security warning issued by Internet Explorer can be safely ignored in this scenario

9. Click Continue To This Website (Not Recommended). The login page for the ATA Console appears as shown in Figure 5-7.

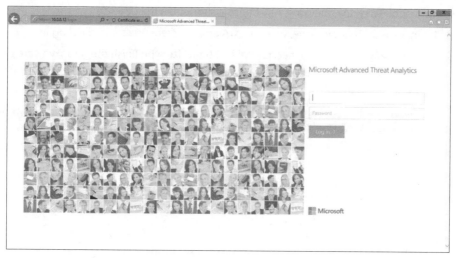

FIGURE 5-7 Microsoft ATA Center login page

At this point, you don't need to log in. The intent here is to show how you can see that the ATA Center installation succeeded. Another validation you can do to ensure that the installation was successfully performed is to verify that the Microsoft Advanced Threat Analytics Center service is running. To verify this, click the Start button, type **services.msc**, and press Enter. You should be able to see the service in a running state, as shown in Figure 5-8.

FIGURE 5-8 Microsoft ATA Center Service running after the installation is finished

> **TIP** If you receive an error during the installation, you can review the error-log files "Microsoft Advanced Threat Analytics Center_<installationdate>.log", located one level above %temp%. To find more information about troubleshooting ATA, look in Appendix A, "Troubleshooting Microsoft Advanced Threat Analytics."

Configuring domain connectivity

Because you can have one or more ATA Gateways installed in your on-premises infrastructure, you need to configure the domain-connectivity settings before installing the ATA Gateway. Use these settings to configure the domain credentials that will be used for the ATA Gateway, which is a read-only user as explained in Chapter 4. Complete the following steps to perform this configuration:

1. Sign in at the ATA Center with the same user account that you used to install it.

2. Double-click the Microsoft ATA Console shortcut in the desktop.

3. In the MS ATA login page, type the ATA admin's credentials and click Log In.

   When this console is opened for the first time after you finish the setup, you are redirected to the ATA Gateway page, where you have the domain-connectivity settings as shown in Figure 5-9.

FIGURE 5-9 Initial domain-connectivity configuration for an ATA Gateway

4. Type the read-only username and password, and type the complete Fully Qualified Domain Name (FQDN) of the domain where the user is located.

5. Click Save.

6. Click the Download ATA Gateway Setup button. Internet Explorer asks you to save a compressed (ZIP) file, as shown in Figure 5-10. This ZIP file contains the ATA Gateway installer and the configuration settings file with the required information to connect to the ATA Center. Click the down arrow beside Save, click Save As, and choose the location where you want to save the file. Click Save, and once the file is downloaded, close the ATA Console.

FIGURE 5-10 A prompt to save the ATA Gateway setup files

Installing ATA Gateway

Before installing the ATA Gateway, you must ensure that all prerequisites for this server role are met. One of the most important settings that must be in place before you start the installation is the network adapter configuration. At this point, you should have at least two network adapters configured: one for communication with the corporate network and the ATA Center (referred to as *Management*), and another one for capturing the traffic (referred to as *Capture*), as shown in Figure 5-11.

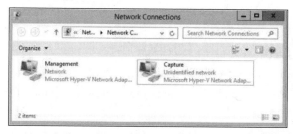

FIGURE 5-11 Network adapters that should be in place prior to installing the ATA Gateway

Another setting that should be adjusted before the ATA Gateway installation is the Domain Name System (DNS) suffix for the Management adapter. You should add the FQDN of the domain in which this server belongs. This setting is automatically configured if the ATA Gateway is domain-joined; however, in this example where the ATA Gateway is not domain-joined, you must add it manually. Figure 5-12 shows an example of how this was done for the Blue Yonder Airlines domain.

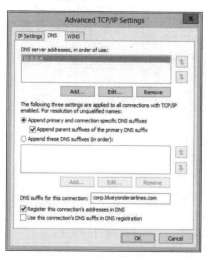

FIGURE 5-12 The DNS suffix for this connection should be the DNS name of the domain for each domain being monitored

When these items are correctly configured and port mirroring is configured between the ATA Gateway and the domain controllers, you can start the installation. (See the "Infrastructure considerations" section in Chapter 4 for more information about port mirroring.)

> **IMPORTANT** Review the supported scenarios for port mirroring at *https://technet.microsoft.com/en-us/library/mt429376.aspx*.

Copy the ZIP file you downloaded in the last step of the previous section, extract it locally on the server, and complete the following steps to start the ATA Gateway installation:

1. Sign in at the ATA Gateway server using an account with local administrative privileges.

2. Open the folder where the ATA Gateway setup files were extracted from, and double-click the Microsoft ATA Gateway Setup.exe file.

3. If the Open File Security Warning dialog box appears, click Run.

4. On the Welcome page, select a language and click Next.

5. On the ATA Gateway Configuration page, you can customize the location where the files will be installed, as shown in Figure 5-13.

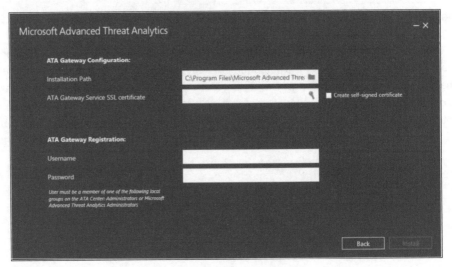

FIGURE 5-13 Customizing the ATA Gateway configuration during the installation

6. Review the following recommendations and issues before making changes to these options:

   - Select the certificate that will be used by the ATA Gateway, or select a self-signed certificate if you are not installing this in a production environment.

   - Review the user's credentials to allow ATA Gateway to register with the ATA Center. This user must be a member of the Administrators group or the Microsoft Advanced Threat Analytics Administrators group on the local machine.

7. When you complete these fields, click Install.

8. The installation progress page appears. When this process is complete, you can click Finish and then click Restart Now to restart the server.

9. After the server restarts, sign in again using the same account you started the installation with. If the Open File Security Warning dialog box appears, click Run. You will notice that the setup will resume and continue the installation.

10. When the installation is complete, you can click Launch to open the ATA Console.

Configuring ATA Gateway

Now that the initial setup is finished, you can start configuring the ATA Gateway. In the ATA Console, type the user's credentials, click Log In, and complete the following steps:

1. Sign in to the ATA Gateway server using an account with local administrator privileges.

2. When you open this console for the first time in the ATA Gateway, you will see the second part of the domain setting configuration, as shown in Figure 5-14.

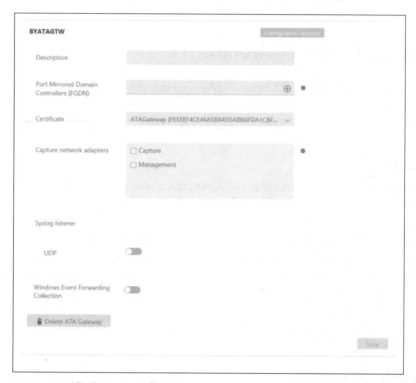

FIGURE 5-14 ATA Gateway configuration

3. Review the following recommendations and issues before making changes to these options:

   - The Description is optional; however, you might use it to describe this server's role or something else.

   - In Port Mirrored Domain Controllers (FQDN), type the FQDN address of one or more domain controllers that will be monitored by the ATA Gateway. After typing an address, click the plus sign (+) to add it.

   - Be aware that the sync is load balanced among all domain controllers in this list.

   - Make sure that the first domain controller in the list is not a read-only domain controller (RODC). You can add an RODC to the list after the initial synchronization is completed. At least one domain controller in the list should be a Global Catalog (GC) server.

For Capture Network Adapter, select the network adapter that has Port Mirroring configured to obtain traffic from the domain controller.

4. Leave the remaining options as they are for now, click Save, and close the console.

> **TIP** You can use the same approach for the ATA Center to validate the installation by verifying whether the service is running. The ATA Gateway error log is also in the same location.

Setting up the ATA environment

After you install the ATA Center and ATA Gateway and make this initial configuration, they will be operating and monitoring the environment. This section explains how to make changes to the default configuration. Some of these changes can vary according to your environment.

Configuring alerts

The monitoring system should alert system administrators about what it is happening, and administrators should review these alerts and take actions to address them. Microsoft ATA automatically sends alerts when it detects a suspicious activity. These alerts can be sent via email or via an alert to your Syslog server (if you configure ATA to use Syslog).

These alerts include a link for system administrators (or whomever receives the notification) to directly view the detected suspicious activity. To configure this option, open the ATA Console (from ATA Center, ATA Gateway, or a workstation—using ATA's FQDN), log in, and complete the following steps:

1. In the ATA Console, select the settings option on the toolbar and click Configuration, as shown in Figure 5-15.

FIGURE 5-15 Settings options toolbar in the ATA Console

2. In the left pane, click Alerts to open the Alert options, as shown in Figure 5-16.

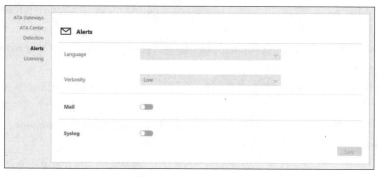

FIGURE 5-16 Customizing alerts

3. Review the following recommendations and issues before making changes to these options:

- The language you select to generate the alerts doesn't influence the language used by the ATA Console.

- By default, the level of log detail (verbosity) for the alert is set to Low. If you decide you need more information, you can change the verbosity to High. Here's an example of how this will affect the amount of information that is sent: in a reconnaissance scenario for account enumeration, with low verbosity ATA sends a notification to the Security Information and Event Management (SIEM) server when the suspicious activity is created. With high verbosity, ATA will send a notification for each account attempted.

4. Blue Yonder Airlines decided to use Mail to send alerts. In this scenario, turn on Mail. You'll see the options shown in Figure 5-17.

FIGURE 5-17 Options available for alerts sent via mail

5. Review the following recommendations and issues before making changes to these options:

- The SMTP Server Endpoint option should have the Mail Exchange (MX) record name for the SMTP Server you will use.

- If this server requires SSL for SMTP connections, ensure that the SSL option is turned on.

- If the SMTP Server requires authentication to allow a message to be sent, turn on Authentication and type the user's credentials.

- In the Send From box, type the email address that represents the user that will be sending the email. In your email server, you can create a read-only, mail-enabled user just to perform this action.

- In the Send To box, add all recipients that should be receiving this email. You can also configure a distribution list in your email server and send this alert to this list.

6. When you have set all the options, click Save.

If your environment has an SIEM server, you can use the same Alert page to enable it. You will need to know the IP address or FQDN name of your SIEM server, as well as the transport protocol (TCP or UDP) and port number to communicate with that server. ATA will send data to the SIEM server using the format specified in Request For Comments (RFCs) 5424 or 3164. All messages sent to SIEM by ATA are formatted using a Common Event Formatting (CEF) standard.

> **MORE INFO** For more information, see RFCs *https://tools.ietf.org/html/rfc5424* and *https://datatracker.ietf.org/doc/rfc3164.*

Monitoring resources

The option that enables system administrators to monitor ATA is called *Health Center*. The Health Center shows potential communication problems between ATA Center and ATA Gateway, such as the example shown in Figure 5-18.

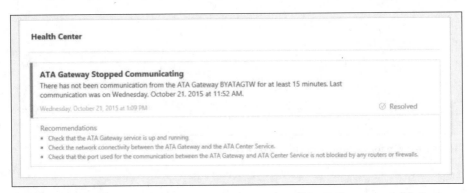

FIGURE 5-18 Alert showing a communication problem between the ATA Gateway and ATA Center

To access the Health Center, click the Health Center icon on the menu bar, as shown in Figure 5-19.

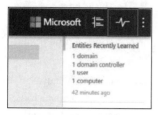

FIGURE 5-19 Access to the Health Center

After you access the Health Center, notice the filters available on the left side of the screen. If you select All, you see all alerts, including the ones that were already resolved. To see which ones are not resolved, click Open. When you resolve an issue and ATA detects that the issue persists, the issue is automatically moved back to the Open Issues list. If ATA detects that an open issue is resolved, it automatically moves it to the Resolved Issues list. The other category you have is Dismissed, which shows issues you do not want ATA to continue to check. In addition to the alert's status, you can visualize the events by using the priority options: High, Medium, and Low.

Detection settings

Different companies have different needs and, therefore, different network behaviors. Some companies might have subnets within their corporate network that are dedicated for testing or validating new apps prior to pushing an app live in production.

The detection settings in ATA allow system administrators to set a list of IP addresses and subnets that have unusual circumstances and should be handled differently than other networks. To customize the detection settings, open the ATA Console, click Configuration in the toolbar, and click Detection in the left pane, as shown in Figure 5-20.

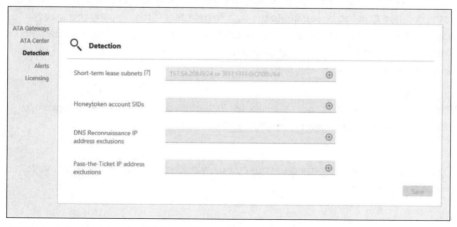

FIGURE 5-20 Customizing the ATA Detection settings

Review the following recommendations and issues before making changes to these options:

- **Short-term lease subnets** Use this option to include subnetworks that have short-term DHCP leases, such as virtual private networks (VPNs) or Wi-Fi. This is particularly important for ATA, because it informs the system that the association between a computer and an IP address from these ranges will have a shorter period of time than it would for other IP addresses in other subnets.

- **Honeytoken account SIDs** *Honeytoken* is an industry term used to specify a user account that should have no network activities. Use this account to trigger a suspicious activity if someone attempts to use this user account. ATA creates a suspicious activity as an indication of malicious activity. In this box, you should add the user account's SID.

> **MORE INFO** You can use the Get-AdUser cmdlet to obtain the user account's SID information. Read *https://technet.microsoft.com/en-us/library/ee617241.aspx* for more information.

- **DNS Reconnaissance IP address exclusions** Use this option to include hosts that belong to your network infrastructure and are authorized to perform DNS reconnaissance. Reconnaissance is the scanning of networks to discover valid information that can be used to map out the environment to assist hackers in their attacks. DNS reconnaissance uses this methodology to discover more information about the DNS servers. This technique is considered an attack, and ATA will trigger an alert if it detects this behavior; however, if you have some workstations in your environment that you use to perform this task, you should add these workstations to this exclusion list. A common scenario for that is when the company has an internal Pen Test (Penetration Test) team. This team can launch a DNS reconnaissance to test their security controls in place.

- **Pass-the-Ticket IP address exclusions** Use this option to include hosts that belong to your network infrastructure and are authorized to perform Pass-the-Ticket. Pass-the-Ticket is a credentials-theft type of attack, in which the attacker steals a user's Kerberos authentication ticket to impersonate the user to gain access to company resources. Again, if you have an internal team that needs to perform this type of attack internally (such as a Pen Test team), you need to add the IP addresses of the hosts that are authorized to perform this test to this list.

> **MORE INFO** For more detail about the Pass-the-Ticket attack, read *https://www.microsoft.com/en-us/download/confirmation.aspx?id=36036*.

After making the changes according to your company's needs, make sure to click Save to commit these changes.

Telemetry settings

By default, ATA collects anonymous telemetry data about usage and transmits this data through a secure channel (over an HTTPS connection) to Microsoft. You can disable this feature in the About window, which can be accessed via the Settings icon (three dots) on the toolbar, as shown in Figure 5-21.

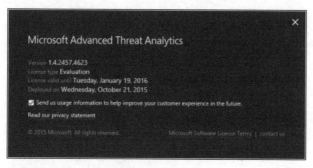

FIGURE 5-21 Changing the telemetry settings

> **TIP** For more information about what type of data Microsoft Telemetry will collect from an ATA environment, visit *https://technet.microsoft.com/en-us/library/mt422979.aspx*.

Database management

ATA stores information in the MongoDB database. By default, the database is located in the ATA Center at %programfiles%\Microsoft Advanced Threat Analytics\Center\MongoDB\bin\ data. If you performed the correct sizing for your environment, you shouldn't be concerned about disk space. However, if you need to move your database to a different drive, you can find a set of documented procedures at TechNet.

> **NOTE** Follow the steps from this article to learn how to move your MongoDB database to another drive or disk: *https://technet.microsoft.com/en-us/library/mt348975.aspx*.

Even though database maintenance might be something you rarely do, troubleshooting scenarios might occur in some instances. For some of these troubleshooting scenarios, you might need to access the database to visualize the records that were committed to the database. In this case, you can use the MongoDB shell commands, or use a Graphical User Interface (GUI) based utility, such as MongoVUE. By default, MongoDB will be listening on port 27017, and you can verify that by accessing the command prompt in the ATA Center and performing the following tasks:

1. Run **netstat –nao**.

2. Take note of the Process ID (PID) bound to the loopback address:process (127.0.0.1:27017).

3. Run **tasklist** to visualize the process that corresponds to the PID.

   You should see that Mongod.exe is the process that is listening on this port, which is the MongoDB service shown in Figure 5-22.

> **NOTE** You can access MongoDB Shell documentation at *https://docs.mongodb.org/ manual/reference/mongo-shell*. For more information about the MongoVUE utility, see *http://www.mongovue.com/*.

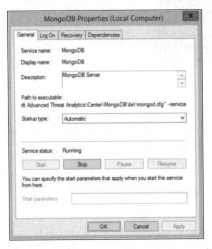

FIGURE 5-22 MongoDB service running on the ATA Center

The information explained previously is important if you need to use utilities such as MongoVUE to access the database, as shown in Figure 5-23.

FIGURE 5-23 MongoVUE connection setting to access the ATA database

Leveraging ATA for threat mitigation and incident response

ATA can help companies mitigate threats by identifying suspicious activities that usually aren't detected until after a security incident has occurred in the environment. Companies that have an incident response policy in place without having a mechanism to stop an attack before it takes place are going to work in a reactive approach for the most part. Although it's important to respond to an incident properly, you also need to identify threats through continuous security monitoring of company resources.

> **MORE INFO** If your company doesn't have a security incident response implemented, make sure to read this article: *https://technet.microsoft.com/en-us/library/cc512623.aspx*.

At this point, all settings are configured and ATA is ready to act. It will start monitoring the environment and trigger alerts in cases where suspicious activities are discovered. In this first release of ATA, the following attacks[1] are identified:

- **Category: Reconnaissance and Brute-Force Suspicious Activities**
  - Reconnaissance using DNS
  - Reconnaissance using Account Enumeration
  - Brute force (LDAP, Kerberos)
- **Category: Identity-Theft Suspicious Activities**
  - Pass-the-Ticket
  - Pass-the-Hash
  - Over-Pass-the-Hash
  - Skeleton Key
  - MS14-068 exploit (Forged PAC)
  - Remote Execution
- **Category: Abnormal-Behavior Suspicious Activities**
  - Abnormal behavior based on resource access, source computers, and work hours (machine-learning algorithm)
  - Massive object deletion
- **Security Issues**
  - Sensitive account exposed in plain text authentication
  - Service exposing accounts in plain text authentication
  - Broken trust
  - Honey Token accounts suspicious activity

---

[1] If you need a terminology glossary for these attacks, visit *https://technet.microsoft.com/en-us/library/mt163704.aspx*.

Real World   How Microsoft ATA detects suspicious activities—under the hood

A TA uses deep-packet inspection to learn and profile the behavior of entities in the network. Each packet sent to the domain controller is analyzed and resolved to the actual entity performing this activity (a user, computer, resource, and other such entity). This information is then used to build an organizational security graph, which represents the relationship between the different entities.

Using this information, ATA can correlate the user, device, resource, and raw network traffic. For example, in a Pass-the-Ticket attack, we use this correlation to identify stolen tickets that were not properly issued to the user by the Active Directory Key Distribution Center.

Once the ATA detection engine has identified the anomaly in the relationships, it will alert the security operator immediately to respond to the alert.

Michael Dubinsky
Senior Program Manager, Microsoft Advance Threat Analytics Team

Reviewing suspicious activities

After deploying ATA, you can use the Timeline option to see a chronological order of events that were detected by ATA. The Timeline option is located on the right side of the Microsoft logo on the toolbar, as shown in Figure 5-24.

FIGURE 5-24 Timeline option located on the toolbar

When you select this option, you see a similar view to the one you saw when you accessed the Health Center. The difference is that, in this case, a timeline is created beside the event to assist you in understanding when such activity took place. Figure 5-25 has an example of a suspicious activity.

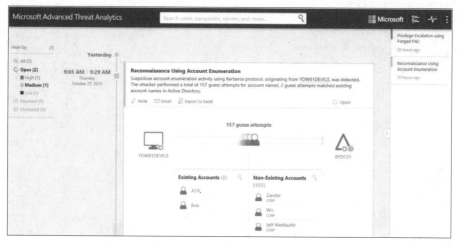

FIGURE 5-25 A reconnaissance attack using account enumeration was detected

The suspicious activity in this case is an account enumeration, which can be considered an attack. The description says: "Suspicious account enumeration activity using Kerberos protocol, originating from YDW81DEVICE, was detected. The attacker performed a total of 157 guess attempts for account names. Two guess attempts matched existing account names in Active Directory." This description shows precisely which device originated the attack (YDW81DEVICE) and what it was able to do. The second part of the alert (which is a medium alert—yellow and visible in Figure 5-26) shows a series of recommendations. In this case, the recommendations are as follows:

- Disconnect YDW81DEVICE from the network, or move it into an isolated environment and start a forensics procedure by investigating the following: unknown processes, services, registry entries, unsigned files, and more.

- Investigate the root cause on YDW81DEVICE.

- Verify that all accounts under "Existing accounts" use a strong password.

You should address these recommendations right away to prevent this device from continuing to try to execute this reconnaissance process. At this point, you might even consider starting an incident response to handle these procedures in the target device.

Attack detection

One of the best ways to reduce the likelihood that attacks will successfully exploit a known vulnerability is to keep all systems up to date using a patch-management process. One of the benefits of ATA is its ability to identify attacks that are trying to exploit known vulnerabilities, as shown in Figure 5-26.

> **MORE INFO** Read this article for more information about patch management and its benefits: *https://msdn.microsoft.com/en-us/library/cc750831.aspx.*

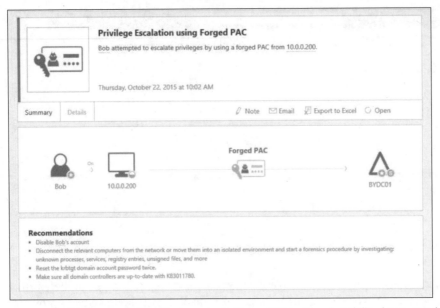

FIGURE 5-26 Privilege escalation attack conducted by exploiting the vulnerability documented in KB3011780

This example shows an attacker using Bob's user account from the host 10.0.0.200 and trying to exploit the vulnerability announced in Microsoft Security Bulletin MS14-068, "Kerberos Key Distribution Center vulnerability using a forged PAC." The two main questions at this point are these:

- Is Bob (a Blue Yonder Airlines full-time employee) really performing this attack, or has someone stolen his identity?

- 10.0.0.200 belongs to the Blue Yonder Airlines corporate network; therefore, this attack was initiated from a workstation that is located internally. Is this workstation compromised, or is it a spoofed address?

To answer these questions, you can follow the recommendations shown at the bottom of the alert, as you can see in Figure 5-26. These recommendations are important and, again, you can start an incident-response process at this point to perform a further investigation.

CHAPTER 6

Introducing Mobile Device Management for Office 365

Organizations of all sizes are adopting a *"bring your own device" (BYOD)* strategy for mobile phones and tablets and embracing the productivity advantages enabled by enterprise mobility solutions. These organizations are also leveraging the power and scalability of cloud-based services, enabling employees to work anywhere and from any device. However, these strategies present IT departments with new challenges in controlling access to corporate data and resources and keeping them protected and secure.

Microsoft Office 365 offers organizations access to several enterprise-level Software as a Service (SaaS) applications and services, such as Exchange and SharePoint Online, and it has established itself as a premium business productivity software provider. Starting in 2015, Microsoft added Mobile Device Management for for Office 365 (MDM for Office 365) to many commercial subscription plans to address the challenges of managing mobile devices and protecting corporate data and resources. With MDM for Office 365, you can manage a diverse mix of mobile devices and set restrictions on how users access corporate information. Additionally, MDM for Office 365 also provides IT and users with the ability to fully reset or selectively wipe data from a device if it is lost, stolen, or reassigned.

This chapter explains how you can use MDM for Office 365 to manage mobile devices and protect corporate resources. It covers the main features and capabilities, discusses design and planning considerations, outlines policy and reporting functionality, and helps you decide if MDM for Office 365 is right for you.

Mobile device management concepts

Organizations have several basic requirements common to any mobile device management strategy or solution:

- Mobile devices, whether company-owned or user-owned, must comply with corporate compliance standards and policies.

- Devices that don't comply with corporate policies must be restricted from accessing company resources until they meet the requirements.

- Company and employee data privacy must be protected.

- Users must have access to resources and services necessary for their productivity.

If these requirements aren't met by a mobile device management solution, IT will have difficulty enforcing company standards and policies, corporate data and resources will be at risk for unauthorized access, and employees will be frustrated and less productive.

In some cases, organizations might need advanced mobile device management features in addition to the basic requirements just mentioned. It is important that you take the time to answer the design and planning considerations covered in Chapter 1, "Understanding Microsoft enterprise mobility solutions," when deciding if advanced mobile device management requirements apply to you. These advanced management requirements might include areas like these:

- Managing mobile applications for different mobile device operating systems

- Integrating cloud-based device management services with existing on-premises device management platforms

- Providing advanced device management capabilities, such as configuring device or networking profiles.

Another consideration in your mobile device management planning might be that your organization already uses device management features included in some on-premises messaging resources. For example, if your organization has existing Microsoft Exchange Servers in your on-premises infrastructure or Exchange Online as part of Office 365, chances are that you're already leveraging the device management features enabled by Exchange ActiveSync. You need to fully understand how Exchange ActiveSync and MDM for Office 365 interact and work together to provide device management features.

Exchange ActiveSync

Exchange ActiveSync (EAS) is a communication protocol used to synchronize messaging platform data with mobile devices and Office 365 Exchange Online and Exchange servers. This synchronization includes information such as email, calendar, and contact data, as well as providing some mobile device management and policy control functionality. EAS also enables native email clients included in most mobile devices to synchronize with Exchange-based user mailboxes and is currently supported on Android, Apple iOS, Microsoft Windows Phone, and Windows 10 Mobile devices.

If you're currently using EAS mobile device management features in Office 365, you're already using Exchange mobile device mailbox policies and the Office 365 Exchange admin center, as shown in Figure 6-1.

FIGURE 6-1 Mobile device management configuration areas in the Exchange admin center in Office 365

EAS provides the following mobile device management functionality:

- Setting mailbox policies to increase device security, such as requiring a minimum device password length and limiting failed password attempts
- Controlling the types of mobile devices that can synchronize with your organization
- Allowing remote wipes that erase all data from a lost or stolen device
- Generating reports for viewing mobile device settings and configuration details

These mobile device management features are valuable, yet they offer only the basic device management functionality needed by most modern enterprise organizations. MDM for Office 365 builds on these mobile device management features and provides you with more control of device security settings, management of Office applications, and access to company data and resources, all without making expensive or complex changes to your on-premises infrastructure or to Exchange Online. If you already use EAS device-management features in Office 365, using MDM for Office 365 isn't a problem. You just set up the policy features, enroll devices, and apply the access and security policies to groups of users. These new policies will override all Exchange ActiveSync mobile device mailbox policies and device access rules in Office 365, making the transition quick and easy.

Mobile Device Management for Office 365

MDM for Office 365 helps you manage security settings and data-access requirements across a wide variety of mobile devices and is included at no extra charge in many popular Office 365 subscriptions. It offers organizations with Office 365 tenants many basic, cost-effective mobile device management features that can be applied to protect all licensed Office 365 users. This is especially valuable to small and medium-sized organizations that don't have an existing on-premises network infrastructure or that don't already use an on-premises device management platform such as Microsoft System Center Configuration Manager. The mobile device management features enabled by MDM for Office 365 greatly expand the management features provided by Exchange ActiveSync in Exchange Online; are tightly integrated with Office 365 productivity apps like Word, Excel, PowerPoint; and help manage access to data stored in Microsoft OneDrive for Business.

> **IMPORTANT** MDM for Office 365 doesn't provide data-access control for resources in on-premises networks. To manage access to on-premises resources and applications, you need a Microsoft Intune subscription. More details about Intune options are covered later in this chapter.

MDM for Office 365 architecture

Microsoft has invested heavily in the seamless integration of different cloud services and MDM for Office 365 is a prime example of this strategy. Additionally, this tight integration of cloud services removes the need for organizations to make costly and complex changes to their on-premises network infrastructure to leverage a broad range of mobile device management and data-access features.

MDM for Office 365 is powered by the Microsoft Intune service, a powerful standalone mobile device management platform and one of the components of Microsoft Enterprise Mobility Suite (EMS). Just as Office 365 leverages Azure Active Directory for identity and user-account management, MDM for Office 365 relies on a subset of the full Intune service functionality to provide mobile device and data-access management. Additionally, MDM for Office 365 is fully integrated with Exchange Online hosted in Office 365, enabling conditional access-management features for Office 365–connected devices and access control for Microsoft Office applications. Figure 6-2 shows the interaction between these services, mobile devices, and managed apps.

This tight architectural integration also enables a transition to Microsoft Intune, either in a standalone configuration or coupled with EMS, for organizations that grow to need more advanced mobile device management features. However, this transition requires coordination with Microsoft Support Services.

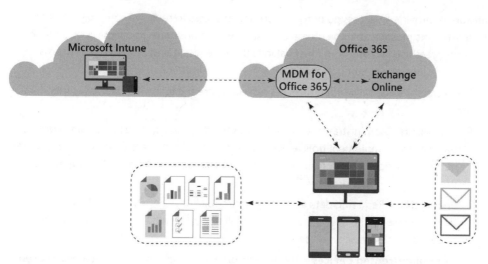

FIGURE 6-2 Office 365 and Microsoft Intune architecture and connectivity for devices and Office applications

IMPORTANT Organizations with Office 365 tenants that exist on their on-premises device management platforms, such as Microsoft System Center Configuration Manager (ConfigMgr), should consider connecting their on-premises platform to Microsoft Intune for mobile device management functionality. MDM for Office 365 doesn't support connections to ConfigMgr or other on-premises device management platforms.

MDM for Office 365 features and capabilities

Building on the mobile device management features offered by Exchange ActiveSync, MDM for Office 365 concentrates its mobile device management features and capabilities in three main areas:

- Configuring device-management settings and restrictions
- Controlling mobile device access to resources
- Protecting data on devices that are lost or stolen

Configuring device management settings and restrictions

The top priorities in mobile device management are protecting mobile devices and preventing unsecured devices from gaining access to company data and resources. These goals are especially important in the modern workplace, where employees often use their personally owned devices to access company resources. Unsecured devices can be compromised by malicious software or provide an access point for unauthorized access to sensitive company information. IT departments must be able to enforce basic security requirements for all the device platforms that employees use, but they also must balance these needs against employees' concerns about overly intrusive or cumbersome device-access requirements.

Organizations using MDM for Office 365 can configure the following device security feature:

- **Passwords**   Administrators can configure security policy settings to require a device password and implement policies such as restricting simple passwords, requiring alphanumeric characters, setting a minimum length, defining the password expiration, and preventing the reuse of passwords.

- **Encryption**   Requiring data encryption on mobile devices helps to protect data from being accessed if the device is lost or stolen. It also prevents the unauthorized recovery of data after the device is erased.

- **Jailbroken/rooted devices**   This feature prevents mobile devices that are jailbroken or rooted from accessing company data and resources. These devices are more susceptible to malware infection, and it's easier for the device's operating system to be compromised.

Controlling mobile device access to resources

If you use Exchange ActiveSync to manage mobile devices in your existing on-premises organization or if you use Exchange Online in Office 365, you might think at this point that MDM for Office 365 really doesn't provide any significant additional benefits. Exchange ActiveSync already provides many of the security features outlined in the previous section. However, to manage access to data in Office 365 and access to Office applications such as Microsoft Word and Excel, you need the richer access-management functionality provided by MDM for Office 365.

With device policies in MDM for Office 365, you can define compliance requirements necessary for enrolled mobile devices to use managed apps to access Office 365 data and resources. You can define several cloud-service, system, application, and device settings using policies:

- **Cloud settings**   You can configure policy settings to block backing up data to cloud services, blocking device synchronization for both documents and photos, and requiring encryption for any permitted backups.

- **System settings**   You can configure policy settings to block device screen captures and to block devices from sending diagnostic data from the device to other services.

- **Application settings**   You can block access to application stores or require a password to access application stores, as well as block video conferencing from a device.

- **Device settings**   You can disable connections to removable storage devices and Bluetooth-enabled devices.

■ **Additional settings**   You can manage other device settings, such as disabling the device camera and certain voice services.

MORE INFO For a complete list of the full capabilities of MDM for Office 365, see "Capabilities of built-in Mobile Device Management for Office 365" at *https://technet.microsoft.com/library/ms.o365.cc.devicepolicysupporteddevice.aspx.*

Protecting data with full and selective wipes

Mobile devices such as smartphones and tablets can contain sensitive company information, and they might have access to your on-premises and Office 365 resources. Additionally, these devices often contain personal information that the device users (or owners in BYOD environments) prefer to keep private. Because both types of data are stored on the mobile device, the company and device user or owner have a vested interest in protecting this information.

MDM for Office 365 has two options for helping to protect your company data from being exposed to or accessed by the wrong people. Office 365 administrators can perform either a remote *selective wipe*, which removes company information and access to company resources from a device, or a remote *full wipe*, which performs a full factory-reset of a device. Users can perform a selective device wipe using the Intune Company Portal app or a full device wipe using their Outlook Web App (OWA) account on Office 365.

These feature options are useful in a variety of common administrative scenarios. The ability to wipe a device can be used when a device is lost or stolen, when retiring devices for replacement, when reassigning devices to new users, or when an employee leaves the company with a personally owned device. You can even configure an MDM for Office 365 policy to automatically wipe a device after a specific number of unsuccessful tries to enter the device's password.

Office 365 administrators can perform either type of remote wipe and monitor the device wipe status from the Office 365 admin center. After you initiate a remote wipe command to a device, the command typically takes around 15 minutes to reach a device connected to the Internet. After the device is wiped, it is removed from the list of managed devices and unenrolled from MDM for Office 365.

Office 365 admin center

The tight architectural integration of MDM for Office 365 within the Office 365 tenant also simplifies deployment and minimizes management overhead for organizations. Instead of having to manage mobile device management platform settings from a separate, standalone product or administration console, you can manage global settings and see the status of enrolled mobile devices right in the Office 365 admin center, as shown in Figure 6-3.

FIGURE 6-3 Office 365 admin center and the mobile-device management home page

You can use the Mobile Device Management for Office 365 console page to do the following:

- **Manage Device Security Policies And Access Rules**   This link directs you to the Device Management section of the Office 365 Compliance Center, where you configure policies for mobile device management.

- **See Device Compliance Report**   This link launches a report to view compliance information about enrolled devices. The reporting specifics are covered in more detail in the "Using the reporting features" section later in this chapter.

- **Select A View**   Using the filters for All or Blocked devices, you can quickly view a summary list of enrolled device information, including the device name, operating system, and operating system version.

- **Manage Settings**   From here, you can see alerts for management requirements that need attention on the main page, as well as configure specific MDM for Office 365 requirements from the Settings page. These settings include configuring domain names in the Office 365 tenant and configuring an Apple Push Notification service (APNs) certificate needed for managing iPhones and iPads that connect to the service. Additionally, you can also configure multi-factor authentication (MFA) from a link to the Azure Active Directory management console and access the Office 365 Compliance Center to configure device security and access policies.

Office 365 Compliance Center

The Office 365 Compliance Center is a centralized management console for all the key compliance-related features for Office 365, such as Exchange and SharePoint Online. You also use the Compliance Center for managing mobile devices with MDM for Office 365. Consolidating management for all the compliance-related functionality in a single area makes settings access easier and simplifies configuring end-to-end compliance requirements.

As shown in Figure 6-4, you'll add security and access policies for devices in the Office 365 Compliance Center, as well as configuring and managing organization-wide device access settings. Additionally, you can view and edit existing security and access policies from this console.

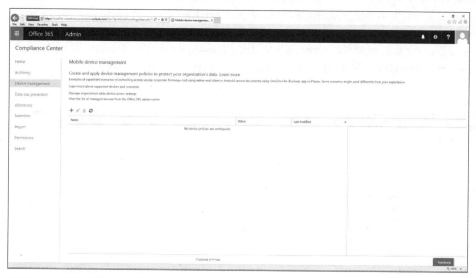

FIGURE 6-4 Office 365 Compliance Center

Planning for MDM for Office 365

Because MDM for Office 365 is included as a feature in the Office 365 cloud service and powered by Microsoft Intune, the prerequisites are simple. No on-premises infrastructure additions are necessary. You just need an Office 365 subscription that includes MDM for Office 365. Currently, MDM for Office 365 is included in the following commercial Office 365 subscriptions:

- Office 365 Business (Business, Essentials, and Premium plans)
- Office 365 Enterprise (E1, E3, and E4 plans)
- Office 365 Education
- Office 365 Government (US Plan 1 & 2, and Government E1, E3, and E4 plans)

All licensed Office 365 users in these plans are eligible for MDM for Office 365 mobile device management features and can enroll a single device.

> **MORE INFO** For the latest information on Office 365 business and enterprise subscription plans and options, see *https://products.office.com/en-us/business/compare-more-office-365-for-business-plans*. For Government subscription plans, see *https://products.office.com/en-us/government/compare-office-365-government-plans*, and for Education subscription plans, see *https://products.office.com/en-us/academic/office-365-education-plan*.

Setting up MDM for Office 365

The first step in setting up MDM for Office 365 is to enable your Office 365 tenant for mobile device management. This is a mostly automated step handled by Office 365 that establishes a connection from the Office 365 tenant to the Microsoft Intune service behind the scenes. You initiate the connection by signing in to your Office 365 tenant as a Global Administrator and navigating to the Mobile Management area in the left navigation pane. From here, you'll see the Set Up Mobile Device Management For Office 365 page, as shown in Figure 6-5.

FIGURE 6-5 The Set Up Mobile Device Management For Office 365 page in the Office 365 admin center

You'll see a summary of the mobile device management features and capabilities provided by MDM for Office 365, as well as links to an overview video and the full MDM for Office 365 document library on Microsoft TechNet. Click the Let's Get Started link to establish the connection with Microsoft Intune. This connection can take several hours to configure, but typically it's completed in less than an hour. Make sure you plan for some delay for this step in your MDM for Office 365 deployment plan. In Chapter 7, "Implementing MDM for Office 365," you'll start from this point in your implementation scenario.

Apple Push Notification service certificate for iOS devices

To manage mobile devices running iOS (iPhones and iPads) with MDM for Office 365, you need an Apple Push Notification service (APNs) certificate. This certificate enables Microsoft Intune to communicate securely with iOS devices over-the-air (OTA) and establishes a partnership with the Apple Push Notification service. MDM for Office 365 and Intune cannot provide or issue this certificate—it must be generated and provided by Apple. Without this certificate, Intune and MDM for Office 365 won't be able to communicate or manage iOS-based devices.

After you initiate the connection between Intune and MDM for Office 365, you'll immediately see a settings alert for an APNs certificate on the main management page, as shown in Figure 6-6. This alert lets you know that you'll need to create and upload an APNs certificate as part of your initial deployment configuration.

FIGURE 6-6 The APNs alert on the Mobile Device Management for Office 365 page in the Office 365 admin center

Completing the APNs certificate configuration process is simple and straightforward. There aren't a lot of complex requirements, and there isn't a lengthy waiting period. You'll just click the Manage Settings link on the Settings alert, and then walk through the following simple steps for configuring an APNs certificate in Office 365:

1. Download a Certificate Signing Request (CSR) from the Office 365 service

2. Create an APNs certificate.

3. Upload the APNs certificate.

> **IMPORTANT** To set up an APNs certificate, you need to use an Apple ID associated with an email account that will remain with your organization, even if the user who manages the account leaves your company. You'll need access to this account to renew the APNs certificate every year.

Adding or configuring a domain

Office 365 can be configured with the default .onmicrosoft.com domain assigned when you subscribe to the service, or you can add and configure a custom domain in the service. Most enterprise-level organizations already have custom domains and will prefer to configure Office 365 to use this custom domain, especially in cases where they are leveraging Exchange Online or configuring Active Directory Federation Services for a single sign-on (SSO) experience. Smaller organizations using Office 365 without a complementary on-premises infrastructure can elect to stick with just the default *<your domain>*.onmicrosoft.com domain.

However, if you're using a custom domain in Office 365 and need to manage Windows Phone or Windows 10 Mobile devices, you must add DNS records for the custom domain at your DNS host. This allows users with email addresses using the custom domain to be properly directed to MDM for Office 365 for device enrollment. If you haven't configured these DNS records, you'll be directed to create Canonical Name (CNAME) records on your host and provided with detailed instructions in the Office 365 domain node in the admin center.

Adding DNS records

After you add the required DNS records, Office 365 users in your organization who sign in on their mobile device using an email address with a custom domain can then be redirected to enroll in MDM for Office 365.

The records you need to add are the following:

```
Host name              Record type Address
EnterpriseEnrollment   CNAME       EnterpriseEnrollment.manage.microsoft.com
EnterpriseRegistration CNAME       EnterpriseRegistration.windows.net
```

You'll need to sign in at your DNS host (for example, eNom.com), go to the section for updating host records for your domain, and then add two new records with the information just shown.

After you add the records and save your changes, wait for the records to update across the Internet. This update can take several hours, but typically finishes in 15–30 minutes. Then go back to the Office 365 admin center, and under Mobile Management, click Manage Settings to complete the setup.

If you don't work with DNS often, it can be a little intimidating to figure out where to add the records at your domain's host or registrar. To help you, Office 365 provides step-by-step guidance for adding DNS records at many popular domain registrars, complete with screen shots of each step.

To see the instructions, find your domain registrar in the list provided in "Create DNS records for Office 365 when you manage your DNS records" (*https://support.office.com/article/b0f3fdca-8a80-4e8e-9ef3-61e8a2a9ab23*), and select the Help link by your registrar's name. (For some registrars, you'll see a link to video step-by-steps as well.) In the Help topic, go to the section for adding CNAME records and follow the steps. Be aware that the steps in the topic provide the values for the basic four CNAME records required for Office 365 (Autodiscover, and so forth), not the MDM record values. So be sure to use the information in the table shown earlier when you're adding the MDM records.

Stacia Snapp
Senior Content Developer, Office 365, Microsoft Corporation

Multi-factor authentication

Multi-factor authentication (MFA) extends the security of cloud-based data and resources beyond using passwords. It helps increase security by requiring users to acknowledge and accept a secondary type of authentication, typically a phone call, text message, or app notification on their mobile device. Users must respond to this secondary authentication request before they can either log in or access information. This type of additional authentication is especially valuable for organizations allowing mobile devices to access company data because the likelihood of these devices being lost or stolen is typically much greater than for traditional desktop workstations or home computers. This strategy also leverages the odds that it's much less likely that the thief of a mobile device also possesses the secondary verification method required by the MFA service. This authentication works by requiring at least two of the following verification methods:

- A password
- A trusted device, like a mobile phone or tablet
- Something unique to you, like biometric data

Office 365 has enabled multi-factor authentication support for administrative roles for several years and has now extended this capability to all licensed Office 365 users. This authentication can be used with applications, such as Office mobile or desktop apps, as well as for accessing data stored in cloud-based services. MFA for Office 365 is powered by Microsoft Azure services and is tightly integrated so that it's easy to use, reliable, and scalable to meet the authentication needs of both small and enterprise-level organizations.

Multi-factor authentication for MDM for Office 365 is managed in the Office 365 admin center, as shown in Figure 6-7.

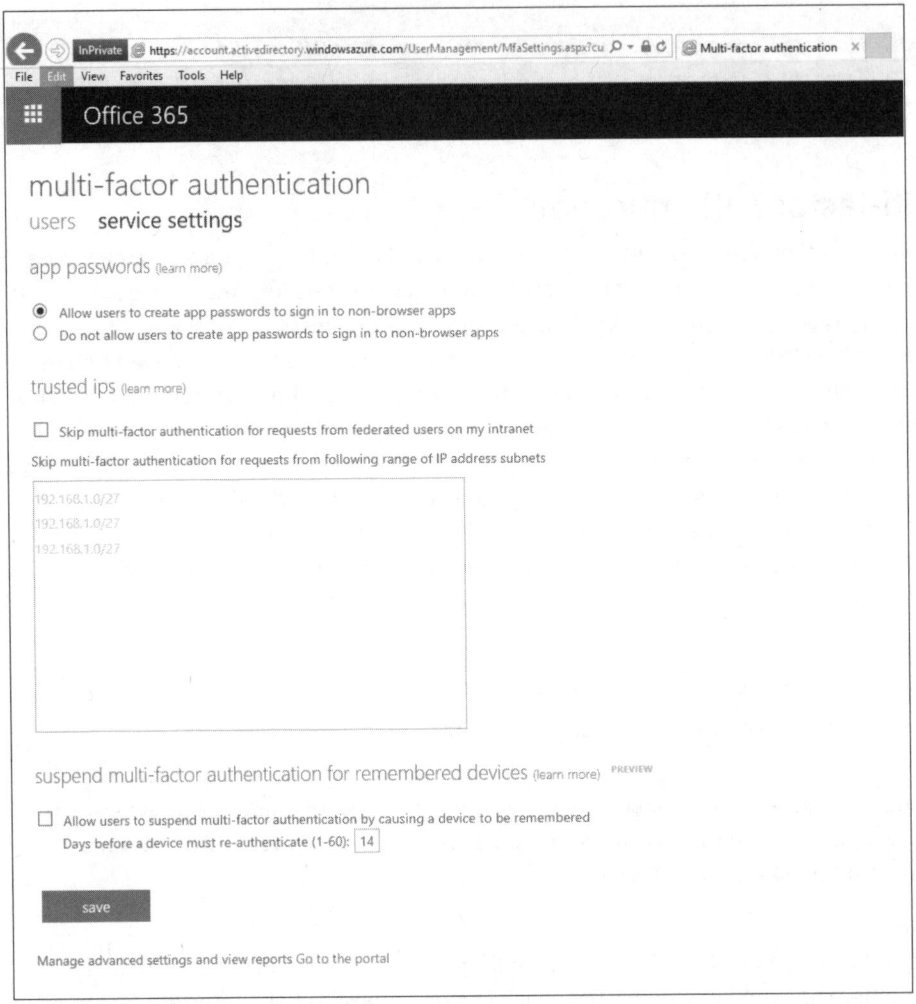

FIGURE 6-7 The Multi-Factor Authentication Service Settings page in the Office 365 admin center

Here you can configure the following service settings that apply to all Office 365 users required to use multi-factor authentication:

- **App passwords**   Allow or restrict users from creating passwords for Office desktop or mobile apps. These passwords are randomly generated by the service.

- **Trusted IPs**   Designate trusted IP address ranges that allow mobile devices using included IP addresses to skip multi-factor authentication when accessing data or resources. This can also include known IP addresses for on-premises resources.

- **Remembered devices**   Allow devices to skip multi-factor authentication requirements for a defined number of days (1 to 60) after they successfully authenticate.

From the Office admin center, you can also manage which Office 365 licensed users are enabled or disabled for multi-factor authentication, and which users have multi-factor authentication enforced but not yet applied. You can select or deselect individual users to set these requirements, or you can bulk update all Office 365 licensed users.

To configure advanced multi-factor authentication features and view reports, you use the Azure Multi-Factor Authentication portal, which is shown in Figure 6-8.

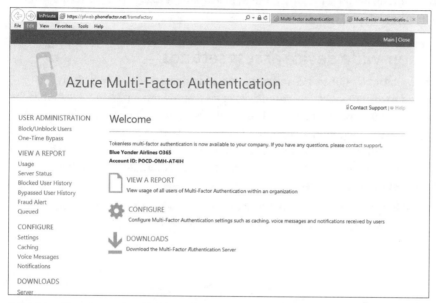

FIGURE 6-8 The advanced multi-factor authentication service portal

> **MORE INFO** To learn more about the requirements and enabling multi-factor authentication in Office 365, see "Plan for multi-factor authentication for Office 365 deployments" at *http://aka.ms/EM2/O365MFA*.

Device management

Creating and configuring policies in MDM for Office 365 is the primary method by which you manage mobile devices and enforce compliance with your organization's standards and policies. You use these device policies to define the security and access requirements that devices (and users) must meet in your organization to connect to Office 365 and use cloud-based resources like Office apps.

The recommended best practice is that you create these policies before people begin enrolling their mobile devices so that the policy requirements are immediately enforced. This approach minimizes gaps in compliance enforcement and allows IT departments to communicate to company employees what they can expect to change about how they use their devices and how they access Office 365.

Remember, before you can create mobile device management policies in Office 365, you must activate and configure MDM for Office 365 in the Office 365 portal. This sets up the connection between Office 365 and Microsoft Intune and prepares the service for managing mobile devices. You must have Office 365 Global Administrator rights for Office 365 to configure policies. This role group includes the Organization Management role, which has access to the Office 365 admin console and Compliance Center.

Organization-wide device access settings

After you initialize MDM for Office 365 and configure additional service requirements such as APNs or multi-factor authentication, you should configure the organization-wide mobile device management settings for Office 365. These settings apply to all devices connecting to Office 365 and support configuring exceptions to organization-wide device requirements. A good way to look at these settings is to distinguish between device *access* and device *compliance*. Organization-wide settings focus on device access, while security policies (discussed later) focus on compliance.

To configure these settings, use the Organization-Wide Device Access Settings page in the Mobile Device Management node in Office 365, as shown in Figure 6-9.

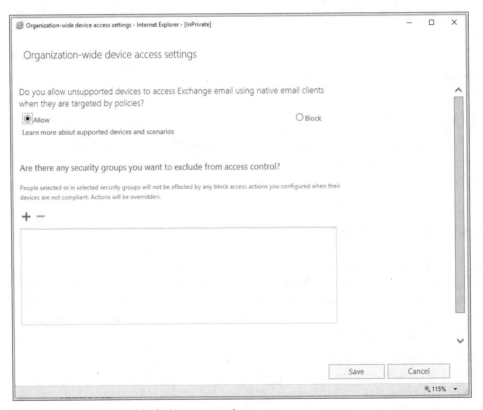

FIGURE 6-9 Organization-wide device access settings

You can use these organization-wide settings to define the following policies:

- **Exchange email access for unsupported devices**   Although MDM for Office 365 supports a broad range of popular widely used mobile devices and operating system platforms, some device platforms and operating system versions aren't supported. For these devices, you can control access to mailboxes hosted on Exchange Online by allowing or blocking access by the native email client application.

- **Excluded users and security groups**   You can exclude individual users or groups of users from access-control policy, even if you configure policies to enforce device-compliance policies.

Security policies

MDM for Office 365 uses security policies to define and enforce requirements for devices enrolled in the service. These policies focus primarily on the state and compliance status of the *device*, not on the identity of the device *user*. This is an important distinction to consider when planning and creating security policies. Remember that the policy defines the security settings you will enforce for mobile devices and is *assigned* to user security groups as part of your mobile device management deployment.

You also need to remember that MDM for Office 365 policies and access rules override all Exchange ActiveSync mobile-device mailbox policies. For example, if you already use Exchange ActiveSync mobile device mailbox policies for Exchange Online and a user synchronizes her mobile device, once you apply an MDM for Office 365 security or access policy to that user (or to a group the user is a member of), the Exchange ActiveSync policies will be ignored. This policy relationship means you can safely delete all Exchange ActiveSync mobile-device mailbox policies after all mobile devices in the organization have been migrated and enrolled in MDM for Office 365.

Security policies in MDM for Office 365 are created and configured in the Mobile Device Management node of the Office 365 Compliance Center. These are groups of settings that control features and compliance requirements on mobile devices. Security policies are easy to create and apply to all supported mobile devices, not just to one mobile device type or platform or another by design. This means you can either create a single security policy to cover all mobile devices or create different security policies for different types of devices (or user groups) in your organization. How you approach this task depends on the mix of mobile devices in your organization and the business needs of your users.

Be sure to keep in mind an important distinction between MDM for Office 365 and Intune security policies. As mentioned, you can apply MDM for Office 365 security policies to devices in your organization, just like you can do with Intune. However, although you can apply MDM for Office 365 security policies to all mobile device, some security settings might not be supported by the device operating system platform. Because the security policies are tailored to specific device platforms with customized settings templates (as they are in Intune), you need to recognize that you might encounter cases where certain policy settings aren't in effect for some devices. This makes it extremely important that you thoroughly understand the differences between various mobile device platforms, versions of their operating systems, and how this affects your security policies.

MDM for Office 365 security policies are organized into the following four sections:

- **General** This section includes defining a name and description for the security policy and is a required section.
- **Access requirements** This is the main policy settings area and contains settings for password requirements, sign-in attempts, encryption, jailbroken status, and what happens if a device doesn't meet the security policy requirements.
- **Configurations** You use this area to block backup, synchronization, store and storage access, and Bluetooth connections.
- **Deployments** This is where you assign the security policy to a user security group. Or you can choose to save the policy and assign it to a group later.

Figure 6-10 shows the access requirement section in the New Device Security Policy Wizard in MDM for Office 365. These requirements can be applied to a user security group (or several groups) later in the wizard, or they can be saved and applied later.

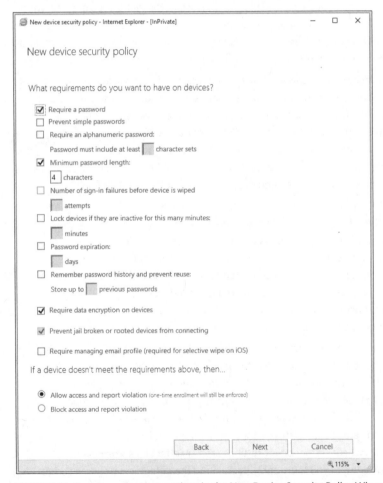

FIGURE 6-10 Access requirement settings in the New Device Security Policy Wizard

Wiping devices

As discussed earlier in this chapter, MDM for Office 365 supports two different ways to wipe a mobile device in cases where the device is lost, stolen, or simply reassigned to another user. This flexibility is one of the key differences between MDM for Office 365 and Exchange Active-Sync mobile-device mailbox policy functionality. Exchange ActiveSync only supports fully wiping a mobile device and returning it to a factory default state. Although this might be preferred in cases where a device is lost or stolen, it's not ideal for BYOD scenarios, where users want to keep personal information and applications on their devices.

Using selective device wipes

Selective device wipes are enabled by the way that MDM for Office 365 compartmentalizes company data and personal information. Each are compartmentalized as completely separate entities so that only the company data and access settings will be removed during a selective wipe command. All personal information, such as personal email, pictures, texts, and contacts, will remain intact on the device after the selective wipe. Administrators can use the Office 365 admin center to perform a selective device wipe, as shown in Figure 6-11. Sign in to Office 365. In the Office 365 admin center, go to Mobile Management, select the device to be wiped, and select Selective Wipe.

FIGURE 6-11 Managed device information and full and selective wipe options

After a selective wipe command is sent to a mobile device, all device policies and Office 365 data hosted by Outlook and One Drive for Business are removed from the device. Depending on the mobile-device operating system platform, the following additional data and applications will also be removed:

- **Android 4+ devices**   Company Portal app, email profiles, and cached email
- **iOS 7.1+ devices**   Company Portal app

IMPORTANT After a selective device wipe (or full device wipe), the device is unenrolled in MDM for Office 365. If the device needs to be managed again, it will need to be re-enrolled before it has access to Office 365 resources.

Using full device wipes

Full device wipes are most commonly used when a device is lost or stolen. It's critical that employees understand that the IT department must be notified as soon as possible after a device goes missing. The longer the gap is between when the device goes missing and when IT is notified, the greater the chance is that sensitive data or access to company resources will be compromised.

To fully wipe a device, administrators log on to the Office 365 admin center, select the device to be wiped, and select Full Wipe. This performs a full data wipe and restores the device to its original factory settings without any user interaction. It deletes all data on the device, including all installed applications, photos, and personal information. Users can also perform a full device wipe from the Options, Mobile Devices node in their Outlook Web App (OWA) account on Office 365, as shown in Figure 6-12.

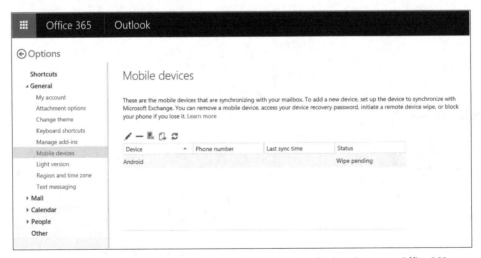

FIGURE 6-12 Full device wipe from the Mobile Devices menu on Outlook Web App on Office 365

Using the reporting features

After you set up security and access policies for your organization and enroll mobile devices in MDM for Office 365, how do you review the compliance status for these devices? It's a great first step to set up mobile device management functionality, but if you don't review the actual status, you'll never know whether devices are being managed as you intended.

MDM for Office 365 shows you information about devices, policies, and compliance status in several ways:

- **Office 365 admin center**   From the Mobile Management node, you can view the device name, device model, platform, platform version, device hardware ID, and the last time the device was synchronized with Office 365 for all managed devices.

- **Office 365 Compliance Center**   From the Device Management node, you can view all security policies, policy status, and when the policy was last modified for all MDM for Office 365 security policies.

- **Device Compliance Report**   From the Office 365 admin center, you can launch the Device Compliance Report. You use this report to review and filter the following mobile device parameters for your Office 365 organization:

  - Total device count

  - Compliant device count

  - Violation reported device count

  - Blocked device count

  - Wiped device count

  - User name

  - Device type

  - First sync time

  - Last sync time

  - Compliance status

Using the Device Compliance Report, which is shown in Figure 6-13, you can quickly see all the devices managed by MDM for Office 365 and the details for the fields mentioned in the preceding list. Using the filtering controls, you can further sort the results by device status or by device operating system:

- **Filter By Status**   You can filter by compliant, reported violation, blocked, or device wiped status.

- **Filter By OS**   You can filter by Android, iOS, Windows Phone, or Other operating systems.

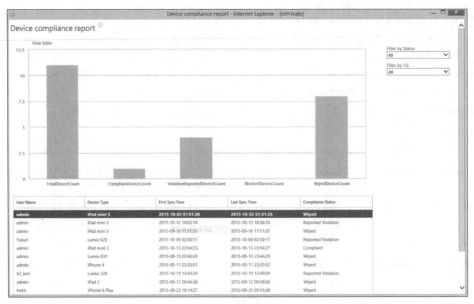

FIGURE 6-13 Device Compliance Report in MDM for Office 365

Choosing MDM for Office 365

This book covers the features and capabilities of MDM for Office 365 and, in Chapter 7, outlines the specific configuration steps for a deployment scenario for a fictitious company, Blue Yonder Airlines. In this scenario, MDM for Office 365 is the best choice for the scenario requirements. The example will show you how to get started using the service. But now that you've read about the service in more depth, you need to determine if MDM for Office 365 is the right fit for your organization and your business needs.

As mentioned earlier in the chapter, MDM for Office 365 is a great choice for some organizations and some scenarios. The features and capabilities offered by the service improve upon the mobile device management capabilities of Exchange ActiveSync, and because it's included at no extra charge in most Office 365 commercial plans, the price is difficult to beat. The fundamental question to answer for your mobile device management planning and design efforts is whether MDM for Office 365 features are comprehensive enough for your needs and your given budget.

Chapter 1 includes a feature comparison chart for Exchange ActiveSync, MDM for Office 365, and Microsoft Intune. It is included here again (in Table 6-1) for ease of reference because you've learned more about mobile application management in Chapter 2, "Introducing mobile application management with Microsoft Intune," and Chapter 3, "Implementing MAM." Since you've learned more about Intune's advanced device and app management features, you're now in a better position to review this feature comparison and choose a solution that best fits your organization's needs before continuing on to Chapter 7.

TABLE 6-1 Comparison of mobility management features for Exchange ActiveSync, MDM for Office 365, and Microsoft Intune

Category	Feature	Exchange ActiveSync	MDM for Office 365	Microsoft Intune
Device configurations	Inventory mobile devices that access corporate applications	x	x	x
	Remote factory reset (full device wipe)	x	x	x
	Mobile device configuration settings (PIN length, PIN required, lock time, and similar)	x	x	x
	Self-service password reset	x	x	x
Basic mobile device and app management	Provides reporting on devices that do not meet IT policy		x	x
	Group-based policies and reporting (the ability to use groups for targeted device configuration)		x	x
	Root and jailbreak detection		x	x
	Remove Office 365 app data from mobile devices while leaving personal data and apps intact (selective wipe)		x	x
	Prevent access to corporate email and documents based upon device enrollment and compliance policies		x	x
Premium mobile device and app management	Self-service company portal for users to enroll their own devices and install corporate apps			x
	App deployment (Android, iOS, Windows Phone, Windows 10)			x
	Deploy certificates, VPN profiles (including app-specific profiles), email profiles, and Wi-Fi profiles			x
	Prevent the cut, copy, paste, and save as operations from being used on data from corporate apps to share the data for use with personal apps (mobile application management)			x
	Secure content viewing via managed browser, PDF viewer, Imager viewer, and AV player apps for Intune			x
	Remote device lock via self-service company portal and via the admin console			x
PC management	Client PC management (for example, Windows 8.1, inventory, antimalware, patch, policies, and similar)			x
	PC software management			x

However, what about scenarios where using MDM for Office 365 makes sense but you still need some of the advanced mobile device management capabilities of Microsoft Intune? As you saw in Chapter 2, with application management, Microsoft is aggressively broadening features and the scope of management possibilities across the entire spectrum of mobile device management. Now Microsoft is offering the ability to combine MDM for Office 365 with the full Microsoft Intune service to offer organizations the best of both worlds.

MDM for Office 365 and Intune coexistence

Just released in late 2015, coexistence functionality provides organizations the flexibility to use the advanced mobile device management features and capabilities for some users in their organization and use MDM for Office 365 to manage other users. Previously, it was an all-or-nothing proposition. You either used MDM for Office 365 (powered in the background by Intune) for mobile device management or Intune for your organization. If you had an Office 365 tenant and also used Intune prior to coexistence, you'd see messages in either service pointing you to the service that had the MDM authority configured to manage mobile devices. Each service would prevent you from using the service that wasn't set as the MDM authority.

Figure 6-14 shows the message displayed in Office 365 if you already configured Intune as the MDM authority for your organization. A similar alert is displayed in the Intune Admin management console indicating that MDM for Office 365 is managing mobile devices if it was configured as the MDM authority.

FIGURE 6-14 Alert in the Office 365 admin center indicating that Microsoft Intune is set to manage all mobile devices

So what does MDM for Office 365 and Intune coexistence really offer you and your organization? In a nutshell, coexistence between these services offers you increased flexibility and another way to structure mobile device management costs. Any single mobile device management solution has difficulty meeting every possible deployment and management scenario for an organization, no matter how well designed or feature rich it is.

Additionally, the modern workplace and the variety of mobile devices is becoming increasingly complex and varied, further complicating management and licensing requirements. There are simply cases where one solution is the best fit for some of your users and another solution is the best fit for other users. MDM for Office 365 and Intune in combination provide the most flexible and most granular mobile device management services to your entire organization, no matter what devices need which management features.

MDM for Office 365 and Intune coexistence management

Configuring Intune and MDM for Office 365 coexistence requires just a few simple steps and offers maximum flexibility for switching the mobile device management authority for these services.

Set up coexistence

To set up coexistence with Intune for your Office 365 tenant, you simply activate both services, using the Office 365 Portal for MDM for Office 365 and the Intune admin console. (See Figure 6-15.) This activation can be completed in any order.

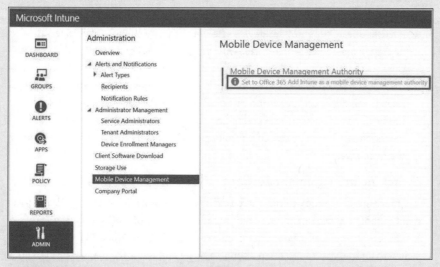

FIGURE 6-15 Setting the mobile device management authority in Microsoft Intune

Which service manages what or who?

When an Office tenant is configured for coexistence, another key decision an IT department must make is which users (and which of their devices) will be managed by which service, and then they must set the corresponding configurations. The answer to the "who is managed by what" question is determined by a new concept called *user-level management authority*. For each user, which service manages what is based on the licenses assigned to the user and the following criteria:

- If the user has only an Office 365 license assigned and does *not* have an Intune license assigned, the user's management authority is "Office 365" and the user's enrolled devices will be managed by MDM for Office 365.

- If the user has an Intune license assigned, regardless of whether the user has Office 365 license assigned or not, the user's management authority is "Intune" and the user's enrolled devices will be managed by Intune.

What happens when switching management authority?

You can easily switch the user-level management authority for a user by changing the license assignment of the user according to the criteria just stated. However, you must be mindful to what happens to enrolled devices when the user's management authority changes:

- If a user unenrolls and then later re-enrolls his device after an authority change has taken place, all existing settings from the previous authority will be removed by the unenrollment, and the new settings from the new authority will be applied to the device's re-enrollment.

- If a user doesn't unenroll his device after an authority change takes place, the next time the device synchronizes with the service (either manually or automatically according to a schedule), the new settings from the new authority overwrite the existing settings that were previously set on the device by the old authority. Keep in mind that an "overwrite" is not a "cleanup and rewrite," so be aware of the settings differences between the two authorities. You should verify that the management authority switch doesn't result in policy-setting conflicts caused by the overwrite.

Switching management authority at the user or tenant level

Another area where coexistence can be extremely helpful is in switching the management authority at the tenant level. Before the availability of coexistence, if you managed users and devices in your organization using MDM for Office 365 and wanted to switch to Intune, you had to open a support ticket and request a manual reset of your tenant's management authority, as well as unenroll and re-enroll all your user devices. This was also a tenant-wide change, meaning that all users and devices had to have the same management authority and all were affected by any change.

With coexistence, now you simply switch user-license settings to the preferred management authority. You can set this by individual user or configure it for all users in your organization. This allows for maximum flexibility, does away with having to open a support ticket, and doesn't require devices to be unenrolled and re-enrolled.

Owen Yen
Senior Program Manager, Enterprise Mobility, Microsoft Corporation

Managing APNs

What about some of the initial mobile device management requirements for managing iOS devices? As discussed earlier, configuring an APNs certificate in MDM for Office 365 was one of the first service-configuration steps needed to manage iPhones and iPads. Both MDM for Office 365 and Microsoft Intune have this requirement, and during a coexistence scenario, there are some important considerations for this tenant-wide setting.

When MDM for Office 365 and Intune are configured for coexistence, you must use the same APNs certificate in both services. You configure this by uploading the APNs certificate from either service, but only once from either service. If different APNs certificates are loaded from MDM for Office 365 and Intune, only the most recent APNs certificate will be used. All iOS devices enrolled with the older APNs certificate will no longer be managed.

Make sure you follow this guidance when you renew your APNs certificate and reload it to either service.

Managing conditional access

Conditional access to Exchange and SharePoint Online can be configured for both MDM for Office 365 and Intune, but there are some fundamental differences between the implementations of each service:

- **Intune**   Conditional access is specifically turned ON or OFF in the Intune Admin Console. This feature, and the corresponding targeting and deployment assignment, can be configured separately for Exchange Online and SharePoint Online.

- **MDM for Office 365**   Conditional access cannot be explicitly assigned; instead, it is implicitly set when an MDM for Office 365 security policy is defined and targeted or deployed to a particular security group. The conditional-access policy is also applied to both Exchange Online and SharePoint Online concurrently. You cannot configure each service separately.

Because MDM for Office 365 and Intune services both use the Office 365 tenant, if conditional-access policies are deployed from both services and overlap users or groups, the last conditional-access policy configured and written "wins" for a user and device. This behavior can result in confusion and unintended information-access scenarios. Make sure when configuring conditional access for both Intune and MDM for Office 365, you target each policy to mutually exclusive user security groups. This approach will help avoid unintentionally overwriting conditional-access-policy settings between the two services.

CHAPTER 7

Implementing Mobile Device Management for Office 365

I n Chapter 6, "Introducing Mobile Device Management for Office 365," you learned how Mobile Device Management for Office 365 (MDM for Office 365) helps you meet the challenges of managing mobile devices and protecting company information, without making changes to your existing on-premises infrastructure. By defining organization-wide access requirements and configuring security policies, you can manage how user devices are enrolled in the service and how they connect to Office 365 services and resources, as well as managing other requirements like device passwords. Remember, security policies define and enforce requirements for devices enrolled in the service, and these policies focus primarily on the state and compliance status of the mobile device.

In this chapter, you'll adopt the persona of the senior enterprise administrator for Blue Yonder Airlines responsible for managing the mobile device management strategy for new acquisitions and remote offices. To meet the mobile device management require-ments described at the end of Chapter 1, "Understanding Microsoft enterprise mobility solutions," you'll deploy MDM for Office 365 for the remote offices, configure access and security policies, enroll mobile devices, and manage device compliance and reporting.

Scenario

As a senior enterprise administrator for Blue Yonder Airlines, you're responsible for planning, designing, and implementing MDM for Office 365 for Blue Yonder's new remote offices that became part of the company in a recent merger. Because of legal requirements established in the merger, the remote offices can't be added to the existing Blue Yonder Airlines network infrastructure right now and must be maintained separately until a later date. Because Blue Yonder Airlines already uses Microsoft Intune to manage most of its employee mobile devices, you're familiar with the fundamentals of enterprise mobility management and know how to use Microsoft solutions for mobile device management.

You know that the new remote offices already use Microsoft Office 365 and host employee mailboxes in Microsoft Exchange Online. Employees in the remote offices use a mix of mobile devices, including Android phones, Apple iPads, iPhones, and Microsoft Windows Phones, most of which are employee owned. They access corporate email mailboxes hosted in Exchange Online from their mobile devices, managed using Exchange ActiveSync policies. However, you need to plan and implement more robust mobile device management features to meet Blue Yonder Airlines' information technology standards, which have more stringent device security and compliance requirements. MDM for Office 365 is a great option for filling this gap until the remote offices can be included in your Intune configuration.

Implementation goals

Building on the existing investment in Office 365 and the mobile management features of Exchange ActiveSync, you now need to address the following implementation goals to successfully complete the MDM for Office 365 rollout for the Blue Yonder Airlines remote offices:

- Require remote-office employees to have managed access to company resources and services from their personal mobile devices and corporate work devices, including remote offices bound by restrictive noncompete agreements

- Enable IT to enforce security, encryption, email, and device policy settings for remote offices using MDM for Office 365

Solution diagram

To meet the MDM for Office 365 implementation goals for this phase of the project, you'll implement the solution shown in Figure 7-1.

> **TIP** This solution diagram provides a high-level overview and basic description of the intended solution architecture. Planning and design considerations for each element of the solution are described in the next section.

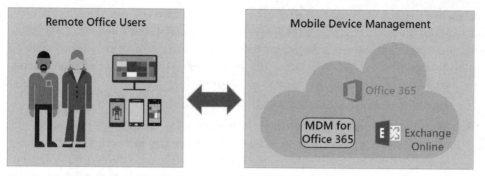

FIGURE 7-1 Using MDM for Office 365 to implement mobile device management to manage remote-office mobile devices and control access to Office 365 resources

This solution is comprised of the following components:

- **Remote-office user** Remote-office users employed by Blue Yonder Airlines need to access company resources, especially corporate email hosted in Exchange Online in Office 365.

- **Mobile device management** MDM for Office 365 will be used to manage mobile devices of remote-office users and ensure that they are in compliance with Blue Yonder Airlines data and resource access policies. These mobile devices will be enrolled in MDM for Office 365, and there aren't any on-premises network infrastructure changes or requirements.

Planning for MDM for Office 365

Now that you've identified your implementation goals and decided on an MDM for Office 365 solution, you need to make sure you understand the planning and design considerations that must be addressed for the deployment to be successful. These considerations include

- Managing user identities
- Creating device security policies
- Supporting required device types

These design considerations were introduced at a high level in Chapter 6, but you should carefully review the information in this section again to make sure you fully understand the impact of your design choices and that you're ready to continue deploying MDM for Office 365.

Identity management

Modern mobile device management solutions really start with and center around user-identity management. How identity is managed and connected is truly the foundation of secure access to company resources, no matter if the resources are located on the premises or in cloud-based services. And because users typically want to be able to work from anywhere and from any mobile device, it's crucial that your identity infrastructure makes it simple and easy for users to reliably access company resources just like they do when in the office. How you configure identity management with MDM for Office 365 will depend on the needs of your organization and the mobile device management solution you deploy.

For organizations using Office 365, there are several options for managing identity:

- **Office 365 accounts (cloud identity)** When you subscribe to Office 365, you automatically have access to Azure Active Directory services and can immediately start creating Office 365–based user accounts from the Users node in the Office 365 admin center. However, creating user accounts in this way doesn't integrate or synchronize these accounts with your existing on-premises directory services, such as accounts hosted in Windows Server Active Directory. This approach might be fine (and even preferable) for smaller organizations without significant on-premises network infrastructure, but it likely isn't the best choice for larger, enterprise-level organizations.

- **On-premises Active Directory accounts (federated identity)** You can connect your existing on-premises Windows Server Active Directory–based accounts with Azure Active Directory services in Office 365 to configure account synchronization and enable single sign-on. Most enterprise-level organizations have made significant investments in their on-premises Active Directory infrastructure and don't want to have to re-create or duplicate user accounts in Office 365, so it makes sense for these organizations to integrate the directories. Synchronizing these directories also enables users to sign in to Office 365–based resources using the same user name and password credentials that they use to access on-premises resources.

Before you get started with MDM for Office 365, you need to decide which identity-management option is best for your organization and how your employees will access resources from their mobile devices. If you choose to configure federated identity with Office 365, you need to complete additional steps and meet additional configuration requirements before configuring access and compliance policies in MDM for Office 365. If you choose to use cloud identity with Office 365, you can start creating user accounts and configuring MDM for Office 365 immediately after the Office 365 service is provisioned. In this chapter, we'll use cloud-identity examples and Office 365–based user accounts only.

> **MORE INFO** For more information about Office 365 user management, see "User Account Management" at *https://technet.microsoft.com/en-us/library/office-365-user-account-management.aspx*. There's also a detailed overview and a step-by-step implementation guide for hybrid (federated) identity with mobile device management, see Chapters 3 and 4 in the *Enterprise Mobility Suite: Managing BYOD and Company-Owned Devices* book from Microsoft Press at *https://aka.ms/EMSdevice/details*.

Policy considerations

Before users enroll mobile devices in MDM for Office 365, you need to carefully plan and configure your access and security policies. Ideally, you'll have all access and security policies configured and assigned to users *before* users enroll their devices. This approach prevents instances where users have enrolled their devices but aren't managed in accordance with your compliance policies. Understanding and correctly configuring these policies will likely require discussions and coordination with personnel in several areas in your organization, including the following:

- Business or department management
- IT administration
- IT security
- Legal

Make sure you plan to consult and coordinate with people in these areas (and any other required areas) as you develop your policy's deployment requirements. If needed, review the planning considerations covered in Chapter 1 before you meet with representatives in these areas. In this chapter, we'll cover the Blue Yonder Airlines requirements as we configure the access and security policies in MDM for Office 365.

Device considerations

Understanding and verifying the types of mobile devices that need to be supported in your organization is also important. It isn't prudent to guess or assume you know what types of devices are in use or will be used in your organization. Reach out to employees and management to learn what devices are being used and how those device types affect your policy requirements. What you discover might amaze you!

For Blue Yonder Airlines, remote-office employees use a wide variety of mobile devices: Android phones, Android tablets, iPhones, iPads, Windows Phones, and Windows tablets. Each of these devices must be able to connect to Blue Yonder Airlines resources in Office 365.

Deploying MDM for Office 365

Now that you've planned for MDM for Office 365 and have a deeper understanding of the needs of Blue Yonder Airlines, you're ready to start implementing your MDM for Office 365 deployment plan. You'll start by preparing the Office 365 tenant and initializing MDM for Office 365 by setting the initial MDM Authority.

Office 365 tenant

The remote offices already have an Office 365 tenant, so you just need to follow the steps in this section to make sure that the tenant supports MDM for Office 365. Remember, Office 365 has several types of subscriptions and not all subscriptions include MDM for Office 365. For a list of the Office 365 subscriptions that include MDM for Office 365, see the "Planning for MDM for Office 365" section in Chapter 6.

Complete the following steps to check your Office 365 subscription:

1. Sign in to the Office 365 admin portal with administrator permissions.

2. In the Office 365 admin center, click Billing and then Subscriptions in the left navigation pane.

3. On the Admin Purchases page (shown in Figure 7-2), verify that the subscription plan listed is the one you wanted (one that includes MDM for Office 365).

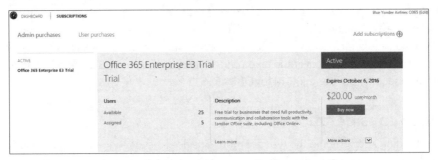

FIGURE 7-2 Office 365 subscription information in the Office 365 admin center

You can also verify that you see the Mobile Management node in the Office 365 admin center left navigation pane. (See Figure 7-3.) If you have an eligible Office 365 subscription that should include MDM for Office 365 and don't see the Mobile Management node, you'll need to contact Microsoft Support to resolve this issue.

Setting the MDM Authority

The next step to get started with MDM for Office 365 is to provision the connection with Microsoft Intune (which actually handles the MDM actions in the background) so that the mobile management features can be configured and assigned to the users at the Blue Yonder Airlines remote offices.

Complete the following steps to provision MDM for Office 365:

1. Continuing in the Office 365 admin center, click Mobile Management in the left navigation pane.

2. Click the Let's Get Started link, shown in Figure 7-3, to start the activation process with Microsoft Intune.

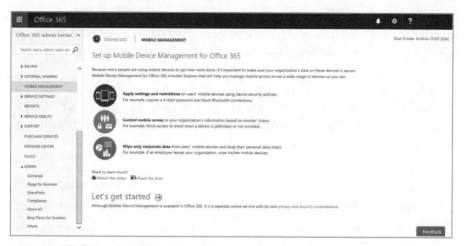

FIGURE 7-3 The link to the Let's Get Started Wizard in the Office 365 admin center

> **TIP** This activation process can take several hours to complete, but typically it takes less than an hour.

After the activation process is completed, you'll see the Mobile Device Management for Office 365 console on the Mobile Management page in the Office 365 admin center.

Configuring MDM for Office 365

Now that you've verified your Office 365 tenant supports MDM for Office 365 and have provisioned the connection with Microsoft Intune, you've completed the prerequisites to successfully start implementing mobile device management for the Blue Yonder Airlines remote offices. By configuring access and security policies (and the policy prerequisites), you'll allow remote-office users to access Office 365 resources and enforce company-information and device-security standards. You'll start by requesting and configuring an Apple Push Notification service (APNs) certificate for iOS-based mobile device management, configure organization-wide access settings, and then configure device-security settings.

Apple Push Notification service certificate for iOS devices

Before iOS-based mobile devices can be enrolled in MDM for Office 365, you need to obtain an APNs certificate and configure it in Office 365. Without this certificate, MDM for Office 365 won't be able to communicate securely with iOS-based mobile devices. Because some Blue Yonder Airlines employees at remote offices use iOS-based mobile devices (iPhones and iPads), you'll need to complete this requirement.

To obtain and install this certificate, download an APNs Certificate Signing Request (CSR) file using the Set Up Mobile Device Management Wizard in the Office 365 admin center and then upload it to the Apple Push Notification Certificates Portal using an Apple ID.

> **IMPORTANT** Remember that this APNs certificate requires renewal on a yearly basis, so make sure you'll have access to the same Apple ID and password used to originally request the APNs certificate. Don't use an Apple ID belonging to someone who might later leave the company. To create an Apple ID, visit *https://appleid.apple.com/account.*

To obtain the Blue Yonder Airlines APNs certificate, you need to complete the following steps:

1. Continuing in the Office 365 admin center, click Mobile Management in the left navigation pane.

2. On the Mobile Device Management For Office 365 page, you should see a Configure APNs For iOS Devices alert in the Settings section of the page. Click Manage Settings to start the Set Up Mobile Device Management Wizard and configure the APNs certificate.

3. In the wizard, next to Create An APNs Certificate For An iOS Device, click Set Up.

4. On the Install Apple Push Notification Certificate, in the Download Certificate Signing Request pane, click Download Your CSR File. Save the .csr file to a location on your computer that you'll remember.

5. Click Next on the Install Apple Push Notification Certificate page.

6. On the Install Apple Push Notification Certificate, in the Create An APNs Certificate pane, click Apple APNs Portal to open the Apple Push Certificates Portal. You'll upload the .cer file to the portal to generate your APNs certificate.

7. Sign in to the Apple APNs Portal with an Apple ID.

8. Click Create A Certificate, and accept the terms of use.

9. On the Create A New Push Certificate page, browse to the Certificate Signing Request you downloaded to your computer from Office 365, and click Upload. You might need to refresh your browser to continue to the next step.

10. Click the Download button, which is shown in Figure 7-4, to save the APNs certificate to your computer. This APNs is usually named something like *MDM_Microsoft Corporation_Certificate.pem.*

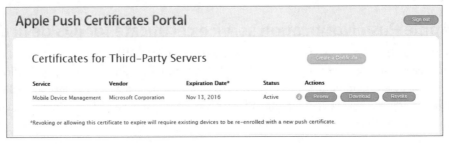

FIGURE 7-4 APNs certificate download page in the Apple Push Certificates Portal

11. Return to Office 365, and on the Install Apple Push Notification Certificate page in the Create An APNs Certificate pane click Next.

12. On the Install Apple Push Notification Certificate page, in the Upload APNs Certificate pane, browse to the APNs certificate you downloaded from the Apple Push Certificates Portal. Select the certificate and then click Open.

13. After you upload the APNs certificate to Office 365, you'll see a message indicating that your APNs was uploaded successfully. Click Finish to complete the APNs process and return to the Office 365 admin center.

14. After you return to the Office 365 admin center, click Manage Settings to verify the status of the APNs certificate. As shown in Figure 7-5, you should see that the Configure an APNs Certificate for iOS Devices set up has completed successfully.

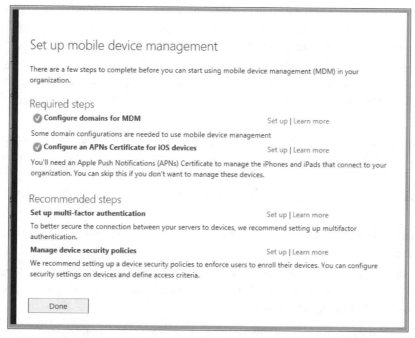

FIGURE 7-5 Successful setup of an APNs certificate in the Setup Mobile Device Management Wizard in Office 365

MORE INFO For the most current guidance for configuring APNs certificates, see "Create an APNs certificate for iOS devices" at *http://aka.ms/EM2/iOSAPN*.

Organization-wide access settings

After you configure the APNs certificate in Office 365, the next step is to configure organization-wide access settings for Blue Yonder Airlines' remote offices. There are two settings you can configure here. The first one lets you allow or block Exchange Online email access for devices that aren't supported by MDM for Office 365. You can also provide a list of security groups that you don't want to be included under MDM for Office 365 access control.

A few things to keep in mind when you're configuring these settings:

- Some devices aren't supported by MDM for Office 365. If any employees have devices that aren't supported, you need to decide if you want to allow or block them from using an Exchange Online account to access Blue Yonder Airlines email. You have to allow all unsupported devices or block all unsupported devices.

- Some mobile device management features in MDM for Office 365 aren't supported for some mobile device operating system platforms. Make sure you understand these exceptions when configuring those policies.

- Decide if any security groups should be excluded from accessing Blue Yonder Airlines Office 365 resources. Users in these security groups will always be able to access Blue Yonder Airlines Office 365 information, even if their mobile devices aren't compliant with the requirements specified in MDM for Office 365 device-security policies.

For Blue Yonder Airlines' remote-office employees, you want to block any devices that aren't supported by MDM for Office 365. You won't add any security groups to be blocked from all access. You'll set up policies for the security groups later in this chapter.

To configure the organization-wide access settings for the remote offices, take the following steps:

1. In the Office 365 admin center, navigate to Mobile Management in the left navigation pane.

2. On the Mobile Device Management For Office 365 page, click Manage Device Security Policies And Access Rules. This will take you to the Mobile Device Management page in the Office 365 Compliance Center.

3. In the Mobile Device Management pane, select Manage Organization-Wide Device Access Settings. The page shown in Figure 7-6 will be displayed.

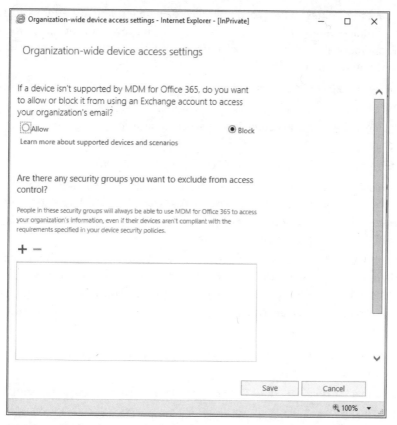

FIGURE 7-6 Configuring organization-wide device access settings in the Office 365 Compliance Center

4. Because you want to increase the level of access security to Blue Yonder Airlines Office corporate email, verify that Block is selected on the Organization-Wide Device Access Settings page. This will prevent noncompliant mobile devices from accessing email hosted on Exchange Online. Additionally, you don't want to exclude any security groups from this policy, so leave the exclusion list blank.

5. Click Save for these updates, and return to the Office 365 Compliance Center.

Security policies

Your next step is to create and deploy security policies for managing mobile devices used at Blue Yonder Airlines remote offices. You'll use these policies to define the security require-ments that devices (and users) must meet in your organization to connect to Office 365 and use cloud-based resources such as Office apps. Mobile devices that don't meet the security-policy requirements will not be able to access Office 365–based resources, including email hosted on Exchange Online. It's a good idea to refer to your security planning documentation as you create these policies because you need to ensure you enforce the same requirements.

Because you have three different types of mobile device operating system platforms in use at the Blue Yonder Airlines remote offices, you'll create a separate device security policy for each platform. This approach allows for more flexibility and is easier to update as new operating system platform versions are released.

Use the following steps to create security policies for Blue Yonder Airlines remote-office users and devices:

1. Continuing in the Office 365 Compliance Center, click Device Management in the left navigation pane.

2. Click the New icon (+) to start the New Device Security Policy Wizard.

3. On the New Device Security Policy page, give the device security policy a name and description. Choosing a friendly policy name makes it easy to find it again later, as well as describing what devices the policy governs. In this scenario, name the policy **Android**, enter a description, and click Next.

4. Next, you'll select the security requirements you want applied to the Android mobile devices used by employees at the Blue Yonder Airlines remote offices. Following the Blue Yonder Airlines security-policy requirements planning documents, configure the following access-requirement settings:

   - Select Require A Password.
   - Ensure that the Minimum Password Length option is selected, and change the character length value to 6.
   - Verify that the default Require Data Encryption On Devices option is selected.
   - Verify that the default Prevent Jail Broken Or Rooted Devices From Connecting option is selected.
   - Click the Block Access And Report Violation option in the If A Device Doesn't Meet The Requirements Above section.

   TIP When creating the device security policy for iOS devices, make sure you select the Require Managing Email Profile option on this page. This is required for supporting selective wipes on iOS devices.

5. Click Next.

6. On the next page, configure the following device connection settings:

   - Select the Require Encrypted Backup option.
   - Select the Block Connection With Removable Storage option.

7. Click Next.

8. Now you can choose to apply the device security policy to the security group by clicking either Yes or No for the "Do you want to apply this policy now?" question. Blue Yonder Airlines has a security group named Android Users that was created to manage remote-office employees using Android devices. To assign this policy to this security group, do the following:

   - Click Yes for the Apply It To One Or More Security Groups option.
   - Click the Add icon (+).
   - Type **Android Users** in the search box and press Enter.
   - Click Add.
   - Click OK.

9. Click Next. On the Review And Confirm The Details page, review the device security-policy configuration settings.

10. Click Finish.

Now you'll see the new device security policy Android in the policy list, and the status will be listed as Turning On. When the policy is configured (typically after just a few minutes), the status will be updated to On. Be sure to create additional device security policies for iOS and Windows Phone mobile devices for the Blue Yonder Airlines remote offices.

On the Mobile Device Management page in the Office 365 Compliance Center (shown in Figure 7-7), you see a list of all the device security settings you configured for Blue Yonder Airlines remote offices. By selecting a policy in the list, you can quickly view the device security-policy details in the Settings pane. Remember, you can update the policy configuration settings by clicking the Edit icon or by double-clicking the security policy.

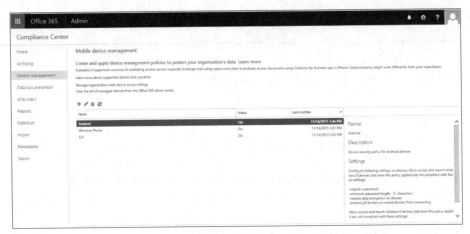

FIGURE 7-7 Device security policies and settings in the Office 365 Compliance Center.

> **MORE INFO** For the most current guidance on creating and deploying device security policies, see "Create and deploy device security policies" at *http://aka.ms/EM2/DevicePolicy.*

Enrolling devices

Now that you've configured the organization-wide access settings, configured device security policies, and assigned the device security policies to users in the Blue Yonder Airlines remote offices, you're ready for employees to enroll their devices into MDM for Office 365. Because you configured access policies to block access to Blue Yonder Airlines Office 365 resources until devices are compliant, employees won't be able to access these resources until they've completed the enrollment process and addressed any compliance-related issues (such as configuring a device access password).

Users can install the Company Portal app and then configure their email account on the mobile device, or they can choose to configure their Blue Yonder Airlines email account on their device and follow the process outlined in the enrollment message. Although both methods will certainly work, many users find that installing the Company Portal app first is simpler and more straightforward. After the enrollment process is completed, they can add their company email account to the device and it immediately is ready to send and receive email messages. For users that already have their Blue Yonder Airlines email configured before a MDM for Office security policy has been assigned to their account, they'll receive the enrollment message and have to enroll their mobile device before they can continue to access their corporate email.

The following sections describe the enrollment process for each of the devices supported in the device policies you've configured for Blue Yonder Airlines:

- Android
- iOS
- Windows Phone

MORE INFO For the most current device enrollment guidance for all operating system platforms, see "Enroll your mobile device in Office 365" at *http://aka.ms/EM2/EnrollDevice.*

Enrolling Android devices

The next step is for remote-office employees using Android devices to enroll in MDM for Office 365 so that they can access their Blue Yonder Airlines email hosted on Exchange Online. Unlike Apple mobile devices, Android devices do not require any additional service-side configuration settings to connect to MDM for Office 365. Android device users should simply install the Microsoft Intune Company Portal app from the Google Play store, or they should follow the enrollment instructions provided in the native Android Mail app message when they try to connect to Office 365.

Users should complete the following steps to enroll their Android devices and access mailboxes hosted on Exchange Online in Office 365:

1. Open the Google Play store on your Android device, and search for Intune Company Portal.
2. Install the Intune Company Portal App and then select Open.
3. When the Company Portal App opens, sign in, and enter your Blue Yonder Airlines account and password.
4. On the Device Enrollment page, select Enroll.
5. Review the settings on the Activate Device Administrator page, and then select Activate.
6. As shown in Figure 7-8, click OK to install your site certificate on the Android mobile device.

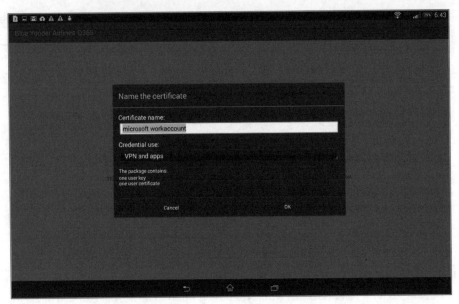

FIGURE 7-8 Certificate acceptance notification on an Android mobile device

7. Depending on the device security policy, you'll typically have to resolve compliance-related issues before the mobile device can be enrolled in MDM for Office 365 as shown in Figure 7-9. For example, you might be prompted to set a device password and encrypt the device. Because you set both of these requirements in the Android device security policy, and the device user is a member of the Android User security group with this policy assigned, this user will need to resolve both of these compliance issues.

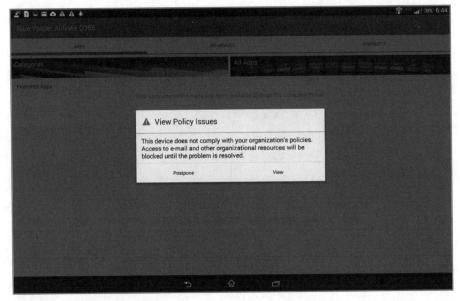

FIGURE 7-9 Policy issues notification on an Android mobile device

8. After the compliance requirements have been resolved, you'll recheck compliance using the Company Portal app by choosing the Check Compliance option. Select My Devices, select the name of your Android device, and then click Check Compliance.

9. After the compliance checks are complete, select Complete On The Company Access Setup page, and then click Done.

Now the remote-office employee has completed the enrollment process, the Android device is compliant with Blue Yonder Airlines device security policy, and the employee can access Blue Yonder Airlines corporate email on their device after configuring an email profile.

Enrolling Apple iOS devices

iPhones and iPads are enrolled into MDM for Office 365 following the same process, and both use the APNs certificate you configured and provisioned earlier. Blue Yonder Airlines remote-office employees can either add their corporate email account to the native iOS mail app or first download and install the Microsoft Intune Company Portal app for iOS from the Apple App Store. The Company Portal app will be installed automatically during the enrollment process outlined by the enrollment email message.

To enroll Apple devices using the enrollment message in the native email profile, device users should complete the following steps:

1. On the iOS device, open the Settings app, click the Mail, Contacts, Calendars section, and choose Add Account.

2. Add the Blue Yonder Airlines Exchange Online account by clicking Exchange, enter your work credentials, click Next, and then click Save.

3. Open the Mail app to synchronize the account with the mobile device.

4. You should see a single email message in your mail account as shown in Figure 7-10. Because you configured the device security policy and assigned the security group to the policy that contains this user, this account won't be synchronized with the Exchange Online mailbox until the device is compliant with the settings defined in the policy. Open the enrollment email message in the Inbox of your email app, and then tap the Get Started Now link.

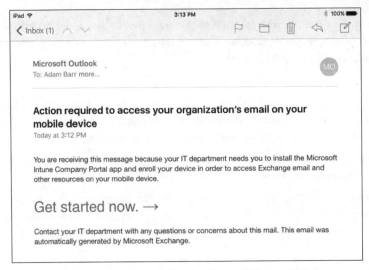

FIGURE 7-10 Intune enrollment email message on an iOS mobile device

5. On the Company Access Setup page, choose View.

6. Click Get and then click Install to install the Company Portal app.

7. Select your Apple ID, and enter your credentials (if required).

8. Open the Company Portal app after it is installed.

9. Sign in with your Blue Yonder Airlines email credentials.

10. Choose Begin on the Company Access Setup page, and then choose Enroll.

11. Click Install on the Install Profile page to install your APNs certificate. You might need to enter your device passcode to continue. Click Install, and on the Mobile Device Management Warning page, click Install to continue when prompted. When the Remote Management dialog box appears, click Trust. When profile installation is complete, click Done. When the Device Enrolled dialog box appears, click OK.

12. Click Continue on the Company Access Setup page. The compliance check might take a few minutes. Depending on the device security policies settings you configured, the user might be prompted to correct compliance-related settings. For example, because you configured a minimum password length requirement of six characters for Blue Yonder Airlines remote employees, these users will need to update the password configuration on the device to meet this requirement and complete the compliance checks successfully.

13. After the compliance checks are complete, click Complete on the Company Access Setup page and then click Done.

As shown in Figure 7-11, the remote-office employee has completed the enrollment process, the iOS device is compliant with the Blue Yonder Airlines device security policy, and the employee has access to Blue Yonder Airlines corporate email on his device.

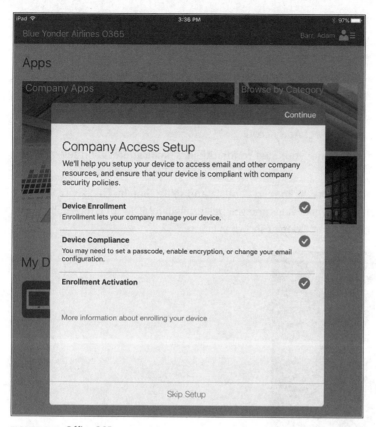

FIGURE 7-11 Office 365 company access setup completed successfully on an iOS mobile device

MORE INFO To help your end users understand how to manage their iOS devices with MDM for Office 365 and Microsoft Intune, you can point them to "Using your iOS device with Intune" at *https://technet.microsoft.com/library/mt598622.aspx.*

Enrolling Windows Phone devices

The process for your users to follow when enrolling Windows Phone devices is similar to the enrollment steps used by both the Android and iOS devices. To enroll Windows Phone devices in MDM for Office 365 for the Blue Yonder Airlines remote-office employees, complete the following steps:

1. On the Windows Phone device, open the Microsoft Store and search for the Microsoft Intune Company Portal app.

2. Install the Company Portal app and then select View. Open the app from the app list. Even though you installed the Company Portal app, you still need to enroll the device in the Intune service before you can access any company resources.

3. Sign in with your Blue Yonder Airlines credentials.

4. Swipe right and select Tap To Enroll or Identify This Device. This takes you to your device Workplace page.

5. On the Workplace page, select Add Account.

6. On the Workplace page, enter your Blue Yonder Airlines email address and select Sign In.

7. Sign in with your Blue Yonder Airlines credentials, and select Sign In. The device will begin checking for compliance issues and will be enrolled in MDM for Office 365. When the process is completed, tap Done as shown in Figure 7-12.

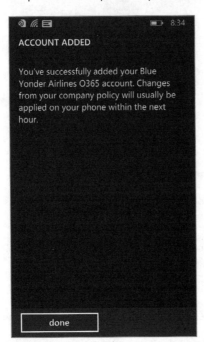

FIGURE 7-12 A Windows Phone mobile device added successfully to MDM for Office 365

8. Although your device is technically enrolled, you won't have access until all the compliance requirements are met. For example, you'll be prompted to update your device password in the Blue Yonder Airlines Windows Phone policy and you will have to set the password before you can access Office 365 resources from the device, as shown in Figure 7-13.

FIGURE 7-13 A security-policy compliance prompt to set a new device password on a Windows Phone mobile device

MORE INFO To help your end users understand how to manage their Windows devices with MDM for Office 365 and Microsoft Intune, you can point them to "Using your Windows device with Intune" at *https://technet.microsoft.com/library/mt427782.aspx*.

Having trouble enrolling your mobile device?

If you're running into issues when you try to enroll a device in MDM for Office 365, try these steps to fix the problem. To start, check the following:

1. Make sure that the device is not already enrolled with another mobile device management provider, such as Intune.

2. Make sure that the device is set to the correct date and time.

3. Switch to a different Wi-Fi or cellular network on the device.

4. For Android or iOS devices, uninstall and reinstall the Intune Company Portal app on the device.

Still stuck? Try these device-specific steps.

For an Android phone or tablet:

1. Make sure the device is running Android 4.0 or later.

2. If you see the error message, "We couldn't enroll this device," sign in to Office 365 and make sure that a license that includes Exchange Online has been assigned to the user who is signed in to the device.

3. Check the Notification Area on the device to see if any required end-user actions are pending, and if they are, complete the actions.

For an iOS phone or tablet:

1. Make sure you create an APNs certificate *(http://aka.ms/EM2/iOSAPN)*.

2. In Settings, General, Profile (or Device Management) on the device, make sure that a Management Profile is not already installed. If it is, remove it.

3. If you see the error message, "Device failed to enroll," sign in to Office 365 and make sure that a license that includes Exchange Online has been assigned to the user who is signed in to the device.

4. If you see the error message, "Profile failed to install," try one of the following:

   • Make sure that Safari is the default browser on the device, and that cookies are not disabled.

   • Reboot the device, and then navigate to *portal.manage.microsoft.com*, sign in with your Office 365 user ID and password, and install the profile manually.

For Windows Phone:

   • Make sure that your domain is set up in Office 365 to work with MDM.

   • Make sure the device is running Windows 8.1 or later.

For Windows RT tablet:

   • Make sure that your domain is set up in Office 365 to work with MDM.

   • Make sure that the user is choosing Turn On rather than choosing Join.

Stacia Snapp
Senior Content Developer, Office 365, Microsoft Corporation

Managing devices

Now that you've enrolled Blue Yonder Airlines remote-office mobile devices, you can manage them using MDM for Office 365. From the Office 365 admin center, you can view enrolled devices using the All or Blocked filter, view device configuration settings, and initiate a full or selective device wipe. Additionally, the Blue Yonder Airlines remote-office employees can manage their devices using the Company Portal app installed on their devices. Let's take a look at some device management features in MDM for Office 365 you have available as an enterprise administrator for Blue Yonder Airlines.

Viewing enrolled devices

To view devices enrolled in MDM for Office 365 as an enterprise administrator for Blue Yonder Airlines, complete the following steps:

1. Log in to Office 365, and navigate to the Office 365 admin center.

2. Click the Mobile Management node in the left navigation pane.

   As shown in Figure 7-14, you can view and sort devices enrolled in MDM for Office 365 by Device Name, Operating System, and Operating System Version. You can also filter the device view by All and Blocked devices. If needed, you can also search the list for devices and manually refresh the device list.

3. When you select an enrolled device in the device view, the device details are displayed on the right side of the device list.

FIGURE 7-14 The Mobile Device Management for Office 365 console in the Office 365 admin center

Viewing the device compliance report

To view the device compliance report in MDM for Office 365 as an enterprise administrator for Blue Yonder Airlines, complete the following steps:

1. Continuing in the Office 365 admin center, click Mobile Management in the left navigation pane.

2. Select See Device Compliance Report. As shown in Figure 7-15, you can view this report to see more detailed information for enrolled devices, including current compliance status, sync status, and totals for different status types for enrolled devices.

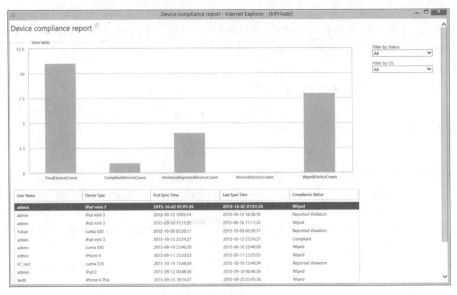

FIGURE 7-15 The MDM for Office 365 device compliance report

Viewing and updating device security policies

To view, manage, or update device security policies as an enterprise administrator for Blue Yonder Airlines, complete the following steps:

1. Continuing in the Office 365 admin center, click Mobile Management in the left navigation pane and select Manage Device Security Policies And Access Rules. This opens the Office 365 Compliance Center.

2. Select a device security policy from the policy list by double-clicking the policy or single-clicking and selecting Edit.

3. As shown in Figure 7-16, update the device security-policy settings and click Save. This saves the policy changes. If you close the policy settings window, you'll see the policy status update to Turning On while the policy is updated. When the update is complete, the policy status will display On.

FIGURE 7-16 Device security-policy settings in the Office 365 Compliance Center

User device management

Managing enrolled devices isn't restricted to enterprise administrators. The Blue Yonder Airlines remote-office employees can also manage their mobile devices using the Company Portal app installed on their devices. Using the app, users can manage the following device areas:

- Rename the device.

- Sync the device with MDM for Office 365.

- Remove the device from management. Removing the device unenrolls the device from MDM for Office 365 and deletes all company data, apps, and data-access permissions.

Depending on the device platform type (Android, iOS, or Windows), the Company Portal app user experience will differ, though each platform will have the same functionality. To manage a mobile device with the Company Portal app as a remote-office employee for Blue Yonder Airlines, complete the following steps:

1. Sign in to your mobile device.

2. Navigate to and open the Company Portal app.

3. Select your device from the device list in the Company Portal to manage the device settings.

4. As shown in Figure 7-17, select the device details setting you want to modify.

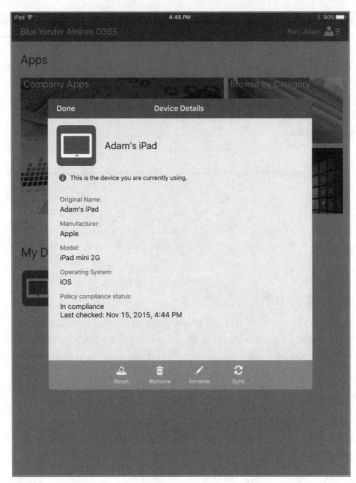

FIGURE 7-17 The device details screen in the Company Portal app on an iOS mobile device

Wiping mobile devices

Mobile devices can be protected if they're lost or stolen by initiating a selective or full device wipe from the Office 365 admin center. This feature is also useful when devices need to be reassigned to a different user. A selective device wipe only removes company data and access settings from the mobile device. All personal information—such as personal email, pictures, texts, and contacts—remain intact on the device. A full device wipe restores the mobile device to its original factory settings and deletes all data on the device, including all installed applications, photos, and personal information. If needed, you can also set up a mobile device management policy that automatically wipes a device after the user unsuccessfully tries to enter the device's password a specific number of times. The mobile device is automatically unenrolled in MDM for Office 365 whenever a selective or full wipe is performed.

Selective device wipe

Now that you've enrolled devices for Blue Yonder Airlines remote-office employees, let's see how you can perform a selective device wipe. Complete the following steps:

1. Sign in to the Office 365 admin portal with administrator permissions.

2. Navigate to the Office 365 admin center and the Manage Mobility node.

3. Select the device you want to selectively wipe from the device list.

4. In the rightmost device settings pane, select Selective Wipe.

5. Read the selective wipe warning, and confirm you want to selectively wipe the device by selecting Yes.

6. After selecting Yes, you'll see the device status change to RetireIssued. When the selective wipe is complete, the device will be removed from the device list and unenrolled in MDM for Office 365.

7. As shown in Figure 7-18, the device owner will be notified of the device unenrollment in the Company Portal app.

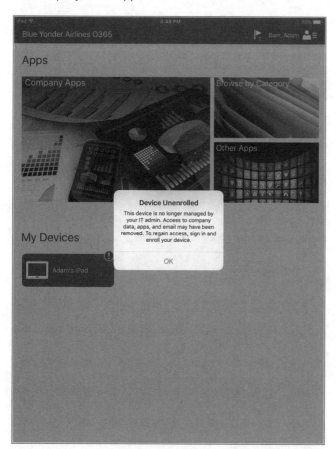

FIGURE 7-18 Device unenrollment message in the Company Portal app on an iOS mobile device

Full device wipe

Now let's see how you can perform a full device wipe. Complete the following steps:

1. Sign in to the Office 365 admin portal with administrator permissions.

2. Navigate to the Office 365 admin center and the Manage Mobility node.

3. Select the device you want to fully wipe from the device list.

4. In the rightmost device settings pane, select Full Wipe.

5. Read the full wipe warning, and confirm you want to fully wipe the device by selecting Yes, as shown in Figure 7-19.

FIGURE 7-19 Full device wipe confirmation message displayed in the Office 365 admin center when wiping a mobile device

6. After selecting Yes, you'll see the device status change to RetireIssued. When the full wipe is complete, the device will be removed from the device list and unenrolled in MDM for Office 365. The device will also be reset to its factory default settings, and all data of the device will be erased.

> **MORE INFO** For the most current selective and full device wipe guidance, see "Wipe a mobile device in Office 365" at *http://aka.ms/EM2/DeviceWipe*.

Troubleshooting Microsoft Advanced Threat Analytics

IN MOST CASES, YOUR Microsoft Advanced Threat Analytics (ATA) implementation was completed smoothly and without incident. Perhaps there were some minor issues you needed to overcome during the initial set up, but everything worked as expected after the deployment was finished. However, sometimes in a production environment where you encounter many variants that are particular to the target network infrastructure, issues may arise not only during the implementation but also during regular operation. This is where ongoing support comes in to perform proper troubleshooting of any problems. As enterprise mobility management features expand across your cloud and on-premises networks, it's crucial that you know how to isolate problems, identify root causes, and quickly resolve issues.

In this appendix, you'll learn key aspects of properly troubleshooting mobility management issues and cover specifics about troubleshooting ATA.

Troubleshooting flow

Troubleshooting an enterprise mobility management issue can be complex, but you can reduce this complexity by leveraging a structured approach to tackle the problem. This section describes the flow of troubleshooting efforts that can help you resolve problems related to the enterprise mobility management products and features presented in scenarios included in this book.

> **NOTE** If you need more information about how to troubleshoot Enterprise Mobility Suite, read Chapter 10 of *Enterprise Mobility Suite: Managing BYOD and Company-Owned Devices* (Microsoft Press, 2015) at *http://aka.ms/EMSdevice/details*.

Initial assessment

The first task for successfully resolving an issue is to define the scope of the problem. You can't start troubleshooting a problem if you can't determine the scope. You have to ask questions and narrow down the scenario.

For example, a typical enterprise mobility scenario is that a user opens a ticket and describes this problem: "I can't access the company's app from my phone." From the troubleshooting perspective, this is a vague statement, so you must ask questions to correctly narrow the scope of the problem. Here are some sample questions to ask:

- When did this problem start?
- Did this work at any point in the past?
- What errors do you receive when you try to access this app?
- Did you recently change anything on this device?
- Does it happen only on this device?
- Where were you connected while trying to access this app (outside the company network or on the premises)?
- If you were outside the company network, were you trying to access this app while connected to Wi-Fi or when using your phone data network?

Depending on the answers of each of these questions, you can summarize the problem and narrow the scope based on the information you gathered. For example:

"User Bob opened an incident on Thursday because he is unable to access Blue Yonder Airlines HR App using his mobile device (Windows Phone 8.1) while connected from his phone data network outside of the corporate network. The issue started on Wednesday night, and the user said he didn't change anything on the device. The user doesn't have another device to try to access the app, but he is able to log on to the corporate network and access other resources."

This is a much more comprehensive scenario description, because you are surfacing a lot of information that can help you eliminate some variables about the issue and give you directions on how to collect data and further narrow the potential root causes. After narrowing the scope, make sure to document all relevant information, such as the following:

- Device operating system and version
- User name
- Domain of the user
- Error messages

Data collection

When troubleshooting hybrid enterprise mobility management scenarios, you must consider where to obtain more detailed information (logs) while the issue is happening. If you collect data to troubleshoot a random issue and the data collection took place while the problem was not happening or in the incorrect log, this data is useless. You'll find it most useful to document issues as they occur in a working scenario. To find the root cause of an issue, you need to ensure that data is collected with tools running in sync with one another.

In scenarios where you need to obtain information from servers located on the premises and the client is coming from the cloud, make sure that the tools are actively collecting data at the same time. In some troubleshooting scenarios, you might have to increase the logging capability on the server, which can have an impact on the server's performance. Make sure to discuss the data-collection plan with the entire team responsible for maintaining the infrastructure and also define a mitigation plan to reduce the likelihood that the server will be negatively affected by this process.

Be aware that data collection for enterprise mobility scenarios also can introduce different challenges when compared to a traditional IT environment, such as these:

- Data-collection procedures and tools in the mobile device will vary according to the vendor.
- Data is always encrypted in transit, which means network traces might not be all that useful.

Data analysis

After obtaining the data from all nodes involved, the first step is to ensure that the data is in sync. Save this data in a secure location with restricted access. Remember that in an enterprise mobility scenario, personal information might be embedded in the logs that were collected. You must keep users' personal data safe during the entire troubleshooting process. Once you have the data, you need to consider the following aspects of the data to be analyzed:

- Do I need to parse this data in order to read it?
- Do I need any special tool to read this data?
- If the amount of data is too big, how can I filter and analyze only what I need?

Answers for these questions will vary according to the data itself. For the technologies that were covered in this book, most of the data can be viewed using built-in tools included with Microsoft Windows, such as Event Viewer and Notepad (to review text logs). Last but not least, understand that the initial data analysis might not indicate the root cause of the issue, but it should give you enough information to help you to build an action plan, which is the next phase.

Action plan

At this point, you should have formed a good hypothesis about why the issue is happening, based on the scope of the problem and the data you analyzed. Based on this hypothesis, you will create an action plan to try to resolve the problem. After building your action plan, make sure to ask yourself the following questions before implementing the plan:

- Will this action plan cause any service interruption (such as restarting the server)?
- Is it necessary to validate this plan in a lab environment before implementing it in production?
- Does this action plan introduce any changes to the production environment, such as installing new software or updates?
- Does this action plan need to be implemented outside of normal business hours?

Based on the answers to these questions, you'll understand what your next step should be. If you identify that this action plan might cause service interruptions, make sure to have a mitigation plan for that. Also, make sure to have a backup plan to roll back the server/workstation or device to the state it was before applying the plan. Last but not least, ensure that the resolution actions are done one by one and not several at a time. Changing multiple options at the same time can make it difficult to know which change really resolves the problem.

> **IMPORTANT** If your action plan fails to resolve the problem, you must go back to the data-collection and data-analysis phases. Evaluate the new data, compare it with the previous data, and verify the differences. Build your new hypothesis for the problem, and create a new action plan. Sometimes this process involves multiple rounds—it all depends on the complexity of the issue.

Validate the behavior and archive the ticket

After implementing the action plan, you need to validate the behavior to see if the issue was resolved. If the issue was resolved, document the effective actions that were taken. Usually after this validation, you will be ready to archive the ticket; however, in some circumstances, you might need to leave it open for monitoring purposes.

If you are dealing with a random issue that occurs only in some circumstances and the user is unable to validate if it was solved, you will need to keep the ticket open. In scenarios like this, you might want to set up the system to collect more data, just in case the issue happens again.

Troubleshooting an ATA installation

Assuming that all prerequisites to install ATA Center and ATA Gateway are in place (review Chapter 4, "*Introducing Microsoft Threat Analytics*," for more information), the installation should be trouble-free. If you are deploying ATA in a physical-switch infrastructure, make sure to follow the recommendations from the switch vendor for creating port mirroring. Usually, the switch vendor also has specific troubleshooting guidelines that vary depending on the switch model.

If you are deploying ATA in a Hyper-V environment, make sure that the domain-controller virtual machine's virtual network port-mirroring mode is configured to be the source, as shown in Figure A-1. In this deployment, the ATA Gateway virtual machine must be configured as the destination.

> **MORE INFO** The following blog post explains in more detail how to configure Hyper-V port mirroring when the traffic you want to capture is from a physical machine (source) and the machine that will do the capturing is a virtual machine (destination): *http://blogs.technet. com/b/networking/archive/2015/01/06/setting-up-port-mirroring-to-capture-mirrored-traffic-on-a-hyper-v-virtual-machine.aspx.*

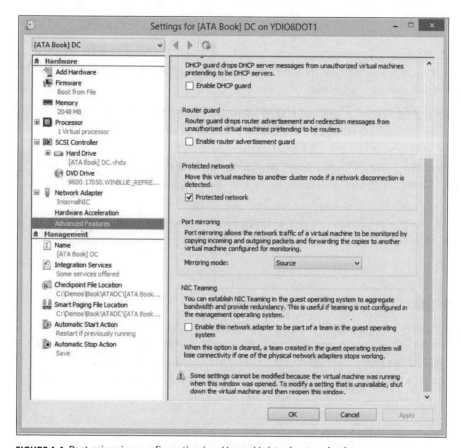

FIGURE A-1 Port-mirroring configuration in a Hyper-V virtual network adapter

In ATA Center, at the end of a successful installation, the log file Microsoft Advanced Threat Analytics Center_*XXXXXXXXX*_1_MsiPackage (where *X* is the date), located one level above

%temp%, will show a summary of the installation. The following indication (in bold) includes confirmation that it was successfully installed:

```
Property(S): INSTALLLEVEL = 1
Property(S): SOURCEDIR = C:\ProgramData\Package Cache\{2EB30FFB-190A-40D7-9522-
FB101E0F2DFD}v1.4.2457.4623\
Property(S): SourcedirProduct = {2EB30FFB-190A-40D7-9522-FB101E0F2DFD}
Property(S): ProductToBeRegistered = 1
MSI (s) (5C:EC) [08:58:59:040]: Note: 1: 1707
```
MSI (s) (5C:EC) [08:58:59:057]: Product: Microsoft Advanced Threat Analytics Center -- Installation completed successfully.

```
MSI (s) (5C:EC) [08:58:59:058]: Windows Installer installed the product. Product Name:
Microsoft Advanced Threat Analytics Center. Product Version: 1.4.2457.4623. Product
Language: 0. Manufacturer: Microsoft. Installation success or error status: 0.
```

If the installation fails, you should open the same log file, go all the way to the end of the file (Ctrl+End) and track back to see where it fails. The other log files that are relevant while troubleshooting installation are these:

- **Microsoft Advanced Threat Analytics Center_*XXXXXXXX*_1_MongoDBPackage** Contains information about the MogoDB installation, which by default is MongoDB 3.0.5 2008R2Plus SSL (64 bit).

- **Microsoft Advanced Threat Analytics Center_*XXXXXXXX*_3_IisUrlRewritePackage** Contains information about URL Rewrite[1] Package installation.

> **IMPORTANT** For ATA Gateway, the relevant logs are Microsoft Advanced Threat Analytics Gateway_*XXXXXXXX*.log and *XXXXXXXXXX*_MsiPackage.log.

Post-installation troubleshooting

After installing ATA Gateway and ATA Center, the first steps are to verify if they are in sync, verify if the services are running, and verify that they can communicate with each other and communicate with the domain controller. If you open ATA Console and you see an error message similar to the one shown in Figure A-2, you need to perform these verifications right away.

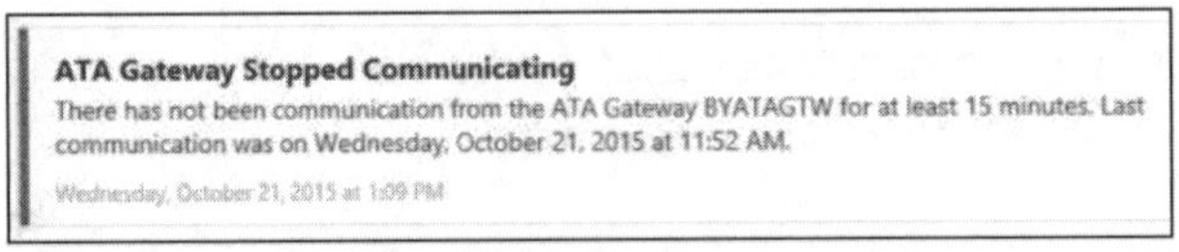

ATA Gateway Stopped Communicating
There has not been communication from the ATA Gateway BYATAGTW for at least 15 minutes. Last communication was on Wednesday, October 21, 2015 at 11:52 AM.

Wednesday, October 21, 2015 at 1:09 PM

FIGURE A-2 An alert showing a communication failure

---

[1] For more information about URL Rewrite, see *http://www.iis.net/downloads/microsoft/url-rewrite*.

The first troubleshooting step in this case is to verify that the Microsoft Advanced Threat Analytics Gateway service is running on ATA Gateway. To do that, click the Windows button on the taskbar, type **services.msc**, and press Enter. If the service is not running, click Start and see if the issue is resolved. If you are unable to start the service and receive an error message similar to the one shown in Figure A-3, you will need to continue troubleshooting.

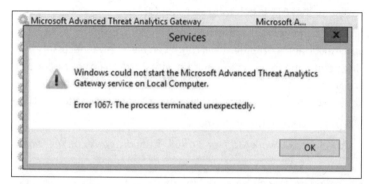

FIGURE A-3 A generic error message that appears when the service is not able to start

This is a post-installation error. In scenarios like this, you need to review the log file located at %programfiles%\Microsoft Advanced Threat Analytics\Gateway\Logs. In this folder, look for a file called Microsoft.Tri.Gateway-Errors.log and open it using a text editor such as Notepad. Search for the keyword *error*, and if you have multiple entries, try to correlate the date and time you received the error message (shown in Figure A-2) with the date and time in the log file. In this case, the problem is shown in the following example in **bold**:

2015-10-21 11:40:45.1095 2288 5   75992705-d0a8-4908-bcfa-de068d7ba6e4 Error [DirectoryServicesClient] Microsoft.Tri.Infrastructure.ExtendedException: **Failed to connect to domain controller [DomainControllerDnsName=bydc01.corp.blueyonderairlines.com] ---> System. DirectoryServices.Protocols.LdapException: The supplied credential is invalid** .

The Microsoft Advanced Threat Analytics Gateway service is not starting because it was unable to connect with the domain controller. Per the error message, the root cause of the problem is related to incorrect credentials.

To resolve this issue, you must re-enter the credentials for the ATA account that was created in Active Directory. Refer to Chapter 5, "Implementing Microsoft Advanced Threat Analytics," for more information about the requirements for this account. Note that error messages in the log file that refer to "credential is invalid" can also appear in cases where the account is expired, locked out, or disabled because of login time restrictions. Make sure to review all these possibilities while troubleshooting this error.

> *TIP* You can also rename the existing error file. If you do that, a new error file will be created automatically the next time the service tries to start. By looking at a new error-log file, you might find it easier to detect where the issue resides.

Troubleshooting ATA operations

After deploying ATA, you start the process of monitoring activities and taking actions based on ATA suggestions. The question that might come up while monitoring alerts on ATA when nothing is happening is, "Is my network really that secure, or is ATA not logging the attacks?" In other words, "How can you validate that ATA is behaving as it should?"

If ATA was correctly installed and the network infrastructure is working properly, ATA is monitoring your network and functioning correctly. It will trigger an alert only if necessary. Understand that if ATA, at any point, is unable to communicate with the domain controller, it won't be able to trigger alerts for attacks. One way to validate the configuration to see if ATA is communicating with the domain controller is by using the Microsoft Network Monitor tool.

Complete the following steps to perform this verification:

1. On the ATA Gateway, download Network Monitor 3.4 x64[2] from *http://aka.ms/netmon3* and install it.

2. Once the installation is complete, launch Network Monitor.

3. On the Start page, select the network interface card that it is used for port mirroring, which is called *Capture* (as it was renamed in Chapter 5). This interface should have an Automatic Private IP Addressing (APIPA) number as shown in Figure A-4.

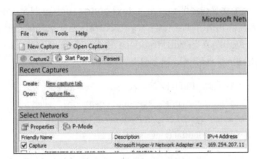

FIGURE A-4 Network Monitor start page with the capture network card selected

4. Click New Capture, and then click Start.

5. Go to a workstation, and sign in with a regular domain account.

6. Go back to ATA Gateway and, on the Network Monitor console, click Stop.

   ATA Gateway should be able to see the authentication traffic between the workstation and the domain controller, similar to the information shown in Figure A-5.

---

[2] Network Monitor 3.4 is not supported on Windows Server 2012; however, for the purpose of this example, it is safe to use. If you need the latest experience in packet analyzes, refer to Microsoft Message Analyzer here: *http://www.microsoft.com/en-us/download/details.aspx?id=44226*.

FIGURE A-5 Network Monitor traffic pattern for a port-mirroring working scenario

> **TIP** If you want to safely simulate an attack to verify if ATA is correctly triggering alerts, follow the instructions in this post: *http://aka.ms/atasimulation*.

In a scenario where ATA isn't working properly, you won't be able to see this traffic. Your first troubleshooting step is to make sure port mirroring[3] is working properly. If you can't see this traffic, it is because there is an issue in this configuration.

Another approach to validate if ATA is working properly is to use Performance Monitor to monitor the following counters:

- ATA Gateway
- NetworkListener Captured Messages / sec
- EntitySender Network Activities
- ATA Center
- EntityReceiver Entity Batch Block Input Items/Sec

> **TIP** For more information about these counters, see the section "Validate Installation" (step 5) in this article: *https://technet.microsoft.com/en-us/library/dn707704.aspx*.

Hardware maintenance

Another troubleshooting scenario you might experience is ATA not triggering new alerts after you made changes to the network interface card—for example, after replacing the card. In this scenario, if you review the log file you might see an error similar to the following:

Error [NetworkListener] Microsoft.Tri.Infrastructure.ExtendedException: **Unavailable network adapters [UnavailableCaptureNetworkAdapterNames=NIC] at Microsoft.Tri.Gateway.Collection.Network. NetworkListener.CreateEtwMessagePusher(IMessageHub messageHub**)

You can see in **bold** in the preceding log that ATA is unable to find the network adapter called *NIC* that's used to capture traffic. In this case, you need to make sure ATA knows which network interface card will be used to capture traffic. Go to ATA Gateway Configuration and, under Capture Network Adapters, select the correct adapter.

---

[3] Here's an example from Juniper Networks on how to troubleshoot port mirroring in its switches:
http://www.juniper.net/documentation/en_US/junos15.1/topics/task/troubleshooting/port-mirroring-qfx-series.html.

Unable to access ATA Console

You need to have access to ATA Console to monitor ATA alerts. If you can't access the console, you won't be able to see what's happening. Not having access to the console doesn't mean ATA is not capturing traffic, because ATA Console is basically a webpage hosted on Internet Information Services (IIS) that displays only captured traffic.

The error message you receive when trying to access the console is the first step to identifying where the troubleshooting process should start. Let's use as an example the error shown in Figure A-6.

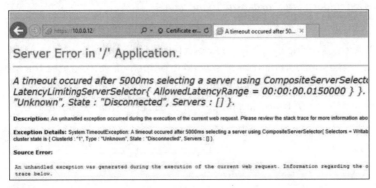

FIGURE A-6 Error message shown while trying to open ATA Console using Internet Explorer

This is a timeout error. In other words, ATA tried to access the website and didn't receive an answer in a timely manner. Complete the following steps to troubleshoot and resolve this error:

1. In ATA Center, open Server Manager, click Tools, and click Internet Information Services (IIS) Manager.

2. On IIS Manager, expand Sites and click the Microsoft ATA Console site.

3. In the Actions pane on the right, under Manage Website, verify that the Start option is unavailable (grayed out). If it is, the website is currently started. Click Restart to restart the website.

4. Try to open ATA Console again.

5. If the issue persists, open Windows PowerShell, type **iisreset**, and press Enter. Wait until you see the output message that says, "Internet services successful restarted."

6. Try to open ATA Console again.

Errors similar to the one shown in Figure A-6 can also indicate that the ATA service is not running. This would also cause problems accessing ATA Console, so make sure you verify that the ATA service is running using the services.msc tool.

Unable to start ATA Center or ATA Gateway

Earlier in this chapter, we gave an example of a scenario where the ATA service was not starting and was presenting the error message shown in Figure A-3. In that case, the problem was caused by ATA not being able to connect with the domain controller. However, if ATA Gateway is not able to access ATA Center, it will also trigger the same generic error message.

To verify the root cause of the issue, you need to review the Microsoft.Tri.Gateway-Errors. log log file. When ATA Gateway is unable to connect with ATA Center, the log will show an error similar to the following:

```
c4ca1680-b50d-4ecd-b179-5720b94faef0 Error [AsyncResult] System.ServiceModel.
EndpointNotFoundException: Could not connect to net.tcp://10.0.0.11:443/
IMonitoringEngine. The connection attempt lasted for a time span of 00:00:21.0133806.
TCP error code 10060: A connection attempt failed because the connected party did not
properly respond after a period of time, or established connection failed because
connected host has failed to respond 10.0.0.11:443.  ---> System.Net.Sockets.
SocketException: A connection attempt failed because the connected party did not properly respond after a
period of time, or established connection failed because connected host has failed to respond 10.0.0.11:443
```

An important part of this log is shown in **bold**. Notice that in this case, ATA Gateway is having exactly the same problem I demonstrated in the previous section when trying to access ATA Console. One possible solution will be to use the same solution applied in the previous section. However, in other scenarios you might face the same error message shown in the log file. One example is when there is a network device (such as a firewall) blocking the communication between ATA Gateway and ATA Center on this port.

Index

A

administration. *See also* Advanced Threat Analytics
(ATA), implementation; also Office 365 Mobile
Device Management (MDM), implementation
access roles, configuring Intune MAM policies, 63–65
APPS workspace, Intune, 24–25, 35–36
location needs, defining, 10
managing applications but not devices, 42–46
Microsoft Intune cloud storage space, 27
Microsoft Intune Software Publisher, 25–27
mobile device life-cycle requirements, 13–14
mobile device management authority, Intune, 22–23
of managed iOS apps, 31–33
Office 365 Mobile Device Management
access settings, 154–155
admin center, 123–124
deployment, 149–151
Intune coexistence with, 140–143, 150–151
Office 365, device management overview, 131–132,
166–170
policy-managed apps (MAM-protected), 36–42
Advanced Threat Analytics (ATA). *See also* Advanced
Threat Analytics (ATA), implementation; also
Advanced Threat Analytics (ATA), troubleshooting
architecture of, 90–91
ATA Center, 86–87. *See also* Advanced Threat Analytics (ATA) Center
ATA Console, 93. *See also* Advanced Threat Analytics
(ATA) Console
ATA Gateway, 86–87. *See also* Advanced Threat Analytics (ATA) Gateway
enterprise security, enhancing, 91
Health Center, resource monitoring, 108–109
in restrictive communication environments, 87–90
infrastructure considerations, 91–92

machine learning and analysis, understanding of,
84–90
overview, 5, 7
protecting on-premises resources, overview, 83–84
standard topology of, 86–87
trial version, 96
Advanced Threat Analytics (ATA) Center, 86–87. *See also*
Advanced Threat Analytics (ATA), troubleshooting
ATA architecture, 90–91
ATA, planning for, 92–93
connection errors, 183
database management, 111–112
in firewall environments, 87–90
installation log file, 177–178
installation of, 97–101
Advanced Threat Analytics (ATA) Console, 93. *See also*
Advanced Threat Analytics (ATA), troubleshooting
alert settings, 106–108
detection settings, 109–110
timeout error message, 182
Advanced Threat Analytics (ATA) Console IIS, 97
Advanced Threat Analytics (ATA) Gateway, 86–87.
See also Advanced Threat Analytics (ATA),
troubleshooting
ATA architecture, 90–91
ATA, planning for, 92–93
configuring, 105–106
connection errors, 183
error log, 106
Health Center, resource monitoring, 108–109
in firewall environments, 87–90
installation, 102–104
log files, 178
Advanced Threat Analytics (ATA), implementation
alerts, configuring, 106–108

185

P

About the authors

YURI DIOGENES is a Senior Content Developer on the CSI Enterprise Mobility Team, focusing on BYOD and Azure Security Center. Previously, Yuri has worked as a writer for the Windows Security Team and as a Support Escalation Engineer for the CSS Forefront Team, also at Microsoft. He has a Master of Science degree in Cybersecurity Intelligence and Forensics from UTICA College and an MBA from FGF in Brazil, and he holds several industry certifications. He is co-author of *Enterprise Mobility Suite: Managing BYOD and Company-Owned Devices* (Microsoft Press, 2015), *Microsoft Forefront Threat Management Gateway (TMG) Administrator's Companion* (Microsoft Press, 2010), and three other *Forefront* titles from Microsoft Press.

JEFF GILBERT is a Senior Content Publishing Manager on the Enterprise Mobility Team at Microsoft. He manages the documentation teams supporting Microsoft System Center Configuration Manager and Microsoft Intune. Prior to returning to management, he was responsible for authoring cross-product solutions to IT business problems involving enterprise client-management technologies, including Microsoft System Center Configuration Manager, Microsoft Intune, and MDOP. Previously, Jeff was the content publishing manager for MDOP and a senior technical writing lead for the Configuration Manager 2007 documentation team. Before joining Microsoft, Jeff was an SMS administrator with the US Army. Jeff is a regular speaker on enterprise client management and MDOP technologies at conferences including the Microsoft Management Summit (MMS), TechEd, IT\Dev Connections, and the Minnesota Management Summit (MMS).

ROBERT MAZZOLI is a Senior Content Developer with Microsoft on the Enterprise Mobility team, working on developing enterprise mobility solutions using the Microsoft Enterprise Mobility Suite and MDM for Office 365. Robert joined the Enterprise Mobility team in 2014 and has been a speaker on enterprise mobility solutions and mobile device management at several conferences, including Ignite 2015 and the 2015 Microsoft MVP Summit. Previously, Robert was a Senior Content Developer for Microsoft Exchange Server and Exchange Online in Office 365, specializing in Exchange hybrid deployments and managing the Exchange Server Deployment Assistant. Before joining Microsoft, Robert owned an information technology consulting business and served as an officer in the United States Navy.

From technical overviews to drilldowns on special topics, get *free* ebooks from Microsoft Press at:

www.microsoftvirtualacademy.com/ebooks

Download your free ebooks in PDF, EPUB, and/or Mobi for Kindle formats.

Look for other great resources at Microsoft Virtual Academy, where you can learn new skills and help advance your career with free Microsoft training delivered by experts.

Microsoft Press

Wait, there's more...

Find more great content and resources in the
Microsoft Press Guided Tours app.

The Microsoft Press Guided Tours app provides
insightful tours by Microsoft Press authors of new and
evolving Microsoft technologies.

- Share text, code, illustrations, videos, and links with peers and friends
- Create and manage highlights and notes
- View resources and download code samples
- Tag resources as favorites or to read later
- Watch explanatory videos
- Copy complete code listings and scripts

Hear about it first.

Get the latest news from Microsoft Press sent to your inbox.

- New and upcoming books

- Special offers

- Free eBooks

- How-to articles

Sign up today at MicrosoftPressStore.com/Newsletters

Microsoft

Visit us today at

microsoftpressstore.com

- **Hundreds of titles available** – Books, eBooks, and online resources from industry experts

- **Free U.S. shipping**

- **eBooks in multiple formats** – Read on your computer, tablet, mobile device, or e-reader

- **Print & eBook Best Value Packs**

- **eBook Deal of the Week** – Save up to 60% on featured titles

- **Newsletter and special offers** – Be the first to hear about new releases, specials, and more

- **Register your book** – Get additional benefits

 Microsoft

Now that you've read the book...

Tell us what you think!

Was it useful?
Did it teach you what you wanted to learn?
Was there room for improvement?

Let us know at http://aka.ms/tellpress

Your feedback goes directly to the staff at Microsoft Press,
and we read every one of your responses. Thanks in advance!

Microsoft